BUS

ALLEN COUNTY PUBLIC LIBRARY

P9-BIG-352

566 W15a
WALLACE, JOSEPH E.
THE AMERICAN MUSEUM OF
NATURAL HISTORY'S BOOK OF

ALLEN COUNTY PUBLIC LIBRARY
FORT WAYNE, INDIANA   46802

You may return this book to any location of
the Allen County Public Library.

DEMCO

THE AMERICAN MUSEUM OF NATURAL HISTORY'S
BOOK OF
DINOSAURS
AND OTHER ANCIENT CREATURES

THE AMERICAN MUSEUM OF NATURAL HISTORY'S
BOOK OF
DINOSAURS
AND OTHER ANCIENT CREATURES

JOSEPH WALLACE

SIMON & SCHUSTER
New York   London   Toronto   Sydney   Tokyo   Singapore

FOREWORD

When completed in the spring of 1996, the American Museum of Natural History's Fossil Halls will drastically reconfigure a cornerstone of the Museum's public image. Prompted by the steady flow of new discoveries and the development of new technologies and approaches to the study of fossils, this major renovation clearly points out that the information in the old exhibition halls that housed the institution's vertebrate fossils was not as timeless as the fascinating specimens themselves.

Rest assured that we here at the Museum are celebrating our heritage by sympathetically restoring the halls to their original architectural grandeur and building the new exhibits around our most popular real fossil skeletons, such as *Tyrannosaurus*, *Apatosaurus*, the mammoth, and the mastodon. The renovation, however, also provides visitors with the most up-to-date information about these venerable heroes of evolutionary history. Through computer interactives, video presentations, and touch-specimens,

we literally bring these long-extinct beasts within reach of the public, providing children and adults with an opportunity both to participate in the latest discoveries and to ponder on-going mysteries.

Our hope is that millions of people will come and share in the exhibition's journey through 500 million years of vertebrate history. It is a journey guaranteed to spark interest—or feed an already existing enthusiasm. In either case, both frequent visitors or long-distance aficionados will find *The American Museum of Natural History's Book of Dinosaurs and Other Ancient Creatures* an entertaining companion—complete with exciting behind-the-scenes details, scientific information, and insider's anecdotes—for exploring the fossil specimens at this esteemed institution.

—**LOWELL DINGUS**, project director of the fossil hall renovation

When all of us were girls and boys
  The greatest of our childhood's joys
Was when our aunts and uncles took us
  Each year to visit Barnum's Circus.
There was the Greatest Aggregation
  Of wonders culled from every nation:
Ferocious beasts whose size gigantic
  Would make you squeal with terror frantic…
But now old Barnum is outclassed,
  A greater showman's come at last.
In Osborn's great menagerie
  A hundred marvels you may see
More strange (according to the labels)
  Than anything in truth or fables.
In Barnum's Circus such a brute as
  The Dog-faced Boy or Guyascutus
He would admit in manner airy
  Was just a bit imaginary.
But Osborn has him skinned a mile.
  Of all the long impressive file
Of giant beasties in his hall
  Not one of them exists at all.

> **—W. D. MATTHEW,**
>   curator in the American Museum
>   of Natural History's Department
>   of Vertebrate Paleontology,
>   circa 1916

THE WEB OF DISCOVERY

With the opening of the American Museum of Natural History's new vertebrate paleontology halls, the Museum is showcasing the fossils of more than eight hundred dinosaurs, mammals, amphibians, fish, and other ancient creatures. Many of these specimens contain hundreds, even thousands of bones, and every bone tells a story.

Some of the stories are apparent to even the most casual observer. One glance at the daggerlike teeth of *Tyrannosaurus rex*, for example, and it's clear that this enormous dinosaur was an eater of meat. And a single look at the magnificent array of mammoths and mastodons on display tells us where our modern-day elephants came from.

Other tales told by the bones can fire our imaginations. *Tyrannosaurus'* newly refigured posture—backbone

Tyrannosaurus versus *Triceratops*: Charles R. Knight's vivid paintings for the American Museum, although not always considered scientifically accurate today, have helped fire the public's love of dinosaurs and other ancient creatures for more than a century.

ABOVE: *Mononykus*, the remarkable ancient bird identified during the Museum's 1990s expeditions to the great Gobi Desert in Mongolia.

BELOW: From the Museum's archives: an irreverent guess as to why *Dimetrodon* and other fossil animals carried sails of skin and bone upon their backs.

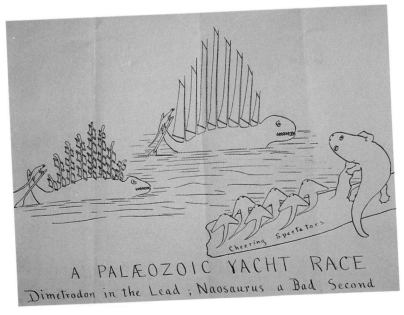

nearly parallel to the ground, head canted forward, tail held stiff behind for balance—allows us to envision an extraordinarily powerful yet agile predator, not the awkward, upright Godzilla we have seen in the past.

Still others give us more subtle hints of interrelationships between animals that might otherwise seem to have little in common. At first glance, the newly discovered *Mononykus* appears to be a small, meat-eating dinosaur. Another look with the help of a more practiced eye, and telltale skeletal features show *Mononykus* to be a unique, seventy-five-million-year-old bird.

In this book, I have included these stories and many others that enable us to envision the dinosaurs and other extinct animals as living creatures. But I have also chosen a different perspective, one that might otherwise elude even the most avid student of dinosaurs and other ancient life. I have tried to describe the men (and, very occasionally, women) who found these countless thousands of bones and who built the remarkable skeletons that have inspired Museum visitors for decades.

These often little-known tales capture the spirit of the Museum's collectors and of the times in which they lived. They allow us to glimpse an aged Barnum Brown, the Museum's premier fossil hunter for more than sixty years, murmuring, "My favorite child," as he introduced visitors to *Tyrannosaurus*. They briefly spotlight *Anatotitan*, undoubtedly the only dinosaur ever traded for a six-gun; *Paraceratherium* (the enormous early rhinoceros found by Roy Chapman Andrews' intrepid team in Mongolia), whose bones had to be protected from bandits; *Apatosaurus*, which wore the wrong head for generations, simply because Henry Fairfield Osborn, the imperious chairman of the Department of Vertebrate Paleontology, wanted it that way; and dozens of others.

What informs all of these anecdotes, as told through unpublished journals, field diaries, letters, decades-old newspaper and magazine articles, and other sources, is the sheer, unadulterated joy of discovery. This joy resounds in Osborn's ecstatic reaction to Brown's letter announcing the 1908 discovery of *Tyrannosaurus* ("[I] feel like a prophet and the son of a prophet"), in Andrews' journals following the 1922 unearthing of *Paraceratherium* ("It was fossil hunting *de luxe* and we laughed & sang for our hearts were light"), and countless other times throughout the history of the collection.

Nor did such emotions thrive only during some bygone, more emotional era. They exist as well in the modern-day paleontologists who continue to add to the Museum's magnificent collection. There is an echo of Andrews' excitement in the tone of Malcolm McKenna's voice as he describes how, after waiting forty years, he finally was able to collect fossils in the Gobi Desert, the first western scientist to do so in more than sixty years. Brown's pleasure in *Tyrannosaurus* is mirrored in John Alexander's face as he gazes at *Notharctus*, the rare North American primate, and in Mark Norell's when he discusses *Mononykus*, the controversial ancient bird.

Those expressions, those tones join together to form a seamless web of discovery that stretches through time over more than a century. This web can never be broken; it will stitch yesterday's and today's discoveries together with those of future generations. It deserves to be commemorated, to be celebrated as the essential human heart behind the American Museum of Natural History's superb collection of bones.

Museum preparators tend to a rogue's gallery of carnosaur heads—actually, a set of casts of the great *T. rex.*

CLASSIFICATIONS

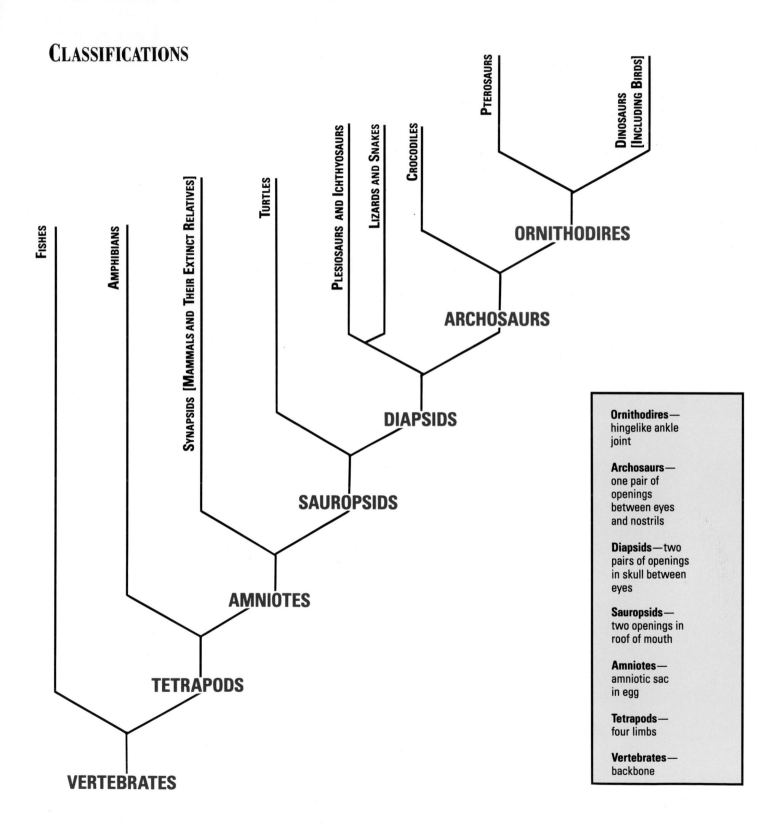

PTEROSAURS

DINOSAURS [INCLUDING BIRDS]

PLESIOSAURS AND ICHTHYOSAURS

LIZARDS AND SNAKES

CROCODILES

SYNAPSIDS [MAMMALS AND THEIR EXTINCT RELATIVES]

TURTLES

FISHES

AMPHIBIANS

ORNITHODIRES

ARCHOSAURS

DIAPSIDS

SAUROPSIDS

AMNIOTES

TETRAPODS

VERTEBRATES

Ornithodires—hingelike ankle joint

Archosaurs—one pair of openings between eyes and nostrils

Diapsids—two pairs of openings in skull between eyes

Sauropsids—two openings in roof of mouth

Amniotes—amniotic sac in egg

Tetrapods—four limbs

Vertebrates—backbone

TIMESCALE

	Period	Events
0	Recent	
10,000	Pleistocene	First modern humans.
1.8 Ma.	Pliocene	First cattle and sheep.
5	Miocene	First apes, mice, and many new mammals.
24	Oligocene	First monkeys, deer, and rhinoceroses.
34	Eocene	First horses, elephants, dogs, and cats.
57	Paleocene	First owls, shrews, and hedgehogs.
65	Cretaceous	First snakes, modern mammals. Dinosaurs die out.
135	Jurassic	First birds. Dinosaurs rule the land.
205	Triassic	First dinosaurs, mammals, turtles, crocodiles, and frogs.
250	Permian	First sail-back reptiles.
290	Carboniferous	First reptiles.
354	Devonian	First amphibians, insects, and spiders.
412	Silurian	First land plants.
435	Ordovician	First nautiloids.
513	Cambrian	First fishes, trilobites, corals, and shellfish.
570		

PART ONE

THE EXPLORERS

HENRY FAIRFIELD OSBORN: THE SIGNATURE OF A GREAT MAN

To do the best work you must live in the period of your research, however remote it may be; marshal the extinct animals before you…revive the physical geography, the temperature, moisture, vegetation, insect life and see before your mind's eye the keen struggle for existence.
Henry Fairfield Osborn, in *Science*, 1901

By 1890, thirteen years after the American Museum of Natural History's heralded opening, its future finally seemed secure. Following years of financial struggle—and the very real chance that the Museum would fold after only a single building

The combination of clutter and opulence of Henry Fairfield Osborn's office and laboratory reflected the personality of the Department of Vertebrate Paleontology's complicated, willful, and brilliant founder.

had been completed—the American Museum's trustees and its president, Morris K. Jesup, finally began to achieve the grand goals set by its founders.

Confident of his plans for the years ahead, Jesup had every reason to be proud of the stabilizing influence he'd brought to the Museum. At the same time, however, he also realized that a serious gap existed in the American Museum's collections, a gap that could leave the institution foundering behind such other ambitious museums as the Peabody Museum in New Haven, Connecticut, and Philadelphia's Academy of Natural Sciences.

The Museum's vaults boasted more than four thousand animal skins, a collection of insects, and an enormous assemblage of fossils of invertebrate animals (those without backbones), but was void of any attraction that would draw visitors to what were in 1890 the remote northwestern outskirts of Manhattan.

Most importantly, the Museum could claim no great dinosaur skeletons, no skulls of ancient mammals, no remains of extinct amphibians or birds. In fact, this imposing institution, which loomed like a fortress over Central Park, contained virtually no fossils of vertebrate animals of any kind.

This glaring lack in the collections was all the more disturbing because during the late 1800s the hunt for vertebrate fossils was a popular and well-publicized pastime. At the time, it seemed as if scientists and amateur bone diggers alike were reporting important new finds nearly every day, steadily adding to the growing body of knowledge of the dinosaurs and other ancient animals.

What must have particularly galled President Jesup was that the two collectors most responsible for the burgeoning national enthusiasm for fossil hunting—Othniel Charles Marsh and Edward Drinker Cope—were both active during the Museum's fledgling years, but neither supplied the institution with specimens. Between the 1860s and the 1890s, Cope and Marsh crisscrossed the country, collecting bones in such vast quantities that it often seemed pure tonnage was

Museum president Morris K. Jesup, who chose Henry Fairfield Osborn to create a new Department of Vertebrate Paleontology.

their only goal. The competition between the two was keen: They raided each other's sites, stole each other's staff, and accused each other of treachery, thievery, and worst of all, scientific inaccuracy.

In short, Cope and Marsh's "Bone Wars" (as the popular press dubbed the feud) was one of the most entertaining spectacles the public had ever witnessed. Jesup, however, looked past the collectors' unprofessional and embarrassing behavior and realized that Cope and Marsh were rapidly accumulating the finest collections of dinosaur fossils ever unearthed. Jesup's assumption was absolutely correct: The two men are credited with identifying more than fifteen hundred genera of ancient vertebrates, an astounding number that included such famous dinosaurs as *Apatosaurus* (then called *Brontosaurus*), *Diplodocus*, *Triceratops*, and *Stegosaurus*.

Yet not a single one of the specimens was destined for the American Museum in New York City. Clearly, Jesup would have to act quickly if the Museum was ever to lay claim to the spectacular dinosaur fossils that were gaining so much public attention.

In the late 1880s, Jesup and the American Museum trustees began searching for the ideal candidate to establish a department of vertebrate paleontology at the Museum. They were looking for someone young, energetic, and well organized—and if possible, someone with sufficient private wealth to help finance some of the workings of the department out of his own pocket.

Henry Fairfield Osborn (1857–1935) met these qualifications, while also boasting a growing reputation for his theories on paleontology and evolution. Although he had not been the Museum's first choice (Jesup had approached Marsh, hoping to get both the collector and his superb fossil collection, but was rebuffed), the trustees liked what they saw in the dark-eyed, intense, self-confident Osborn. And the young paleontologist, at the time a faculty member at Princeton University, made no secret of his interest in creating an important new department at the Museum.

In 1891, after months of planning and negotiation, Osborn, at the age of thirty-four, was named the American Museum's first curator of vertebrate paleontology. In addition, he became a member of the faculty at Columbia University, an arrangement worked out between the Museum and Columbia that still exists today. During the next four decades, first as a curator and then as Museum president, Osborn rarely allowed the public spotlight to shift from him and the paleontology department he cared for so passionately.

Within a decade of his arrival at the Museum, Osborn had gathered perhaps the most talented group of paleontologists ever assembled at a single institution. William Diller (W. D.) Matthew, Walter Granger, Barnum Brown, Albert "Bill" Thomson, Jacob L. Wortman, James W. Gidley, and others all came aboard before 1900. Each of these men possessed the qualities that distinguish only the finest fossil hunters. They were ready and willing to spend months living in primitive conditions in the harshest environments, and once there, they had an unerring instinct (which Osborn considered almost otherworldly) for locating the richest fossil deposits.

Having gathered this highly talented group, Osborn quickly put them to work, establishing an extraordinarily ambitious schedule of field expeditions. Between 1897 and 1902 alone, American Museum parties prospected in eight states in the United States and the southern reaches of South America, with paleontologists returning to many of the richest sites several times.

Henry Fairfield Osborn in an atypically informal pose. In his more than forty years at the Museum, he only occasionally ventured into the field to visit his fossil hunters' research sites.

Whether they were exploring the badlands of Montana or Wyoming, the deserts of Texas, or even the great steppes of Patagonia, the Museum's collectors felt they were journeying back into the past that so fascinated them. Their letters are filled with references to "visiting the Jurassic" and "working in the Cretaceous," as if they were hunting living creatures, not fossils. Without exception, they followed Osborn's dictum to "live in the period of your research."

Back at the Museum, Osborn also knew the importance of hiring brilliant preparators—the skilled technicians whose job it is to remove fossils from surrounding rock, clean and prepare them, and mount them for display. As Osborn immediately realized, funds for future fieldwork depended on the accomplishments of such early preparators as Adam Hermann, Charles Christman, and Charles and Otto Falkenbach. Without their efforts, the results of the fieldwork would never have been seen by the public, and the Museum administration simply would not have continued to support the curator's grandiose plans for expansion of the department.

Perhaps just as importantly, Osborn was willing to dig deep into his own pockets to obtain his goals. When he assumed his position as curator, he agreed to accept no salary and to contribute fifteen hundred dollars each year for departmental expenses. By the end of his tenure, Osborn had spent at least eighty thousand dollars of his own money to build the Museum's collection of fossils and develop the vertebrate paleontology halls into some of the finest in the world.

Osborn also had wealthy friends who took an interest in the Museum—none more prominent or more generous than the financier J. P. Morgan. Serving as Museum treasurer in the 1880s (and helping keep the Museum afloat with personal contributions), Morgan turned his attention to the Department of Vertebrate Paleontology in the nineties and soon became its most dependable benefactor.

During the first two decades the department was in existence, Morgan contributed toward the purchase of Edward Drinker Cope's magnificent fossil vertebrate collection; donated thirty thousand dollars to purchase and restore the Warren Collection of fossils, which included the famous Warren Mastodon; established a research fund for the department; and of crucial importance, donated specially designed railroad freight cars, which enabled field paleontologists to ship tens of thousands of pounds of fossils back to the Museum virtually free of charge.

Only a man with a remarkably strong personality could have accomplished what Osborn did in his early years as curator. A sense

ABOVE: The first completed Museum building (circa 1880), a pioneering structure rising bravely from the swampy outskirts of Manhattan.   *LEFT:* By 1922, the Museum had assumed a more familiar configuration—although the Roosevelt Wing and Hayden Planetarium still remained to be built.

of his powerful character comes through in his writings, even those dating from the dawn of his tenure. Scrupulously polite, his letters and articles gave credit where credit was due, but at the same time, his writing is suffused with absolute certainty in his beliefs. He never doubted that his choices were always the right ones and that those who disagreed with him must be misguided.

George Gaylord Simpson, a curator in the Department of Vertebrate Paleontology for many years, came to the Museum in the 1920s. His memories of Osborn, as recounted in his 1978 memoir, *Concession to the Improbable*, are amused and affectionate, but they also provide a hint of the qualities that enabled Osborn to make himself into a central figure in North American paleontology.

One day, Simpson recalled, Osborn, who by then was president of the Museum, arrived bearing a recently published scientific monograph he had written. Inscribing it to the young curator, he signed, "With the appreciation and best wishes of Henry Fairfield Osborn."

"As this was in heavy, flowing black ink, I reached for a blotter," Simpson wrote, "but he stayed my hand and said, 'Never blot the signature of a great man.'"

Osborn's ego would eventually diminish his reputation. He brooked no criticism on his research on ancient mammals, which was erratic, and he began to focus increasingly on repugnant theories of evolution and race. As a younger man, however, charged with the creation of a new department within an equally young museum, his instincts were sure and his choices nearly always correct.

ABOVE: A youthful Henry Fairfield Osborn (at right) with premier fossil hunter Barnum Brown at Como Bluff, Wyoming, in 1897. *LEFT:* A typical field expedition team during Osborn's reign: Relaxed, confident, and determined to make the American Museum's fossil collection the finest on earth.

BARNUM BROWN: THE SIXTY-SIX-YEAR REIGN OF "MR. BONES"

*I am deeply interested in Barnum Brown's letter
and reports....He is certainly a wonderful man
whom nothing can stop.*

Henry Fairfield Osborn, 1923

When Henry Fairfield Osborn hired Barnum Brown as a staff member of the Department of Vertebrate Paleontology in 1897, he took on an unproven, inexperienced young man who had long been entranced by fossils. Growing up on a Kansas farm with fields filled with ancient shells, Brown (1873–1963) had begun collecting fossils as a boy. Then, as a student at the University of Kansas, he had studied with Samuel Williston, one of Othniel Charles Marsh's chief collectors, and had decided to pursue fossil hunting as a profession.

"This skull alone is worth the summer's work for it is perfect," Barnum Brown wrote to Henry Fairfield Osborn in 1908, after finding *Tyrannosaurus rex*, the Museum's most renowned dinosaur.

Brown (standing, second from left), with preparators Jeremiah Walsh, Peter Kaisen, and Otto Falkenbach, restoring the twenty-thousand-piece jigsaw puzzle of the ankylosaur *Hoplitosaurus*.

Osborn gave Brown a trial as a part-time assistant on an 1896 collecting expedition and then brought him on staff the next summer at a salary of fifty dollars a month. It turned out to be one of the wisest decisions Osborn ever made. As for Brown, he never looked back. His reputation growing in tandem with that of the Museum's, Brown eventually became so famous that his discoveries were covered on the front pages of newspapers across the country. More than sixty years later, "Mr. Bones" was still finding fossils, and until his death in 1963, he could be seen strolling through the halls of the Museum, visiting his prize finds. His list of discoveries is astounding and includes a lion's share of the Museum's most important dinosaurs, most notably *Tyrannosaurus*, *Triceratops*, *Saurolophus*, and *Ankylosaurus*.

Brown did whatever was necessary to secure these specimens for the Museum. He floated a raft down the rugged Red Deer River in Alberta, becoming the first paleontologist to explore the area's extraordinarily rich Late Cretaceous bone beds. He flew twenty thousand miles over North America in a rickety airplane, searching for dinosaur fossils from the air. He dove for fossils in a flooded spring in

Cuba and hunted for ancient mammals in the wilds of Patagonia. He even risked bandits and the bubonic plague and almost died of malaria in India. By the time his illustrious career had ended, he had given so many lectures, appeared on so many radio shows, and written so many articles about his adventures that countless young fossil hunters dreamed of following in his footsteps.

Brown accomplished all this while maintaining the elegant dress, upright posture, and remote, bemused manner of a duke forced to associate with commoners. In all their years together at the Museum, George Gaylord Simpson wrote, his relationship with Brown "never went beyond a courteous but distant acquaintance to become friendship." Many others had the same experience.

Yet Brown's letters from the field to Osborn and others back at the Museum constitute his most personal writing and show a very different sort of man. Far more revealing than any carefully composed lecture or well thought-out magazine article, these letters reveal his love for what he did and his liking for the people who shared his passions.

Throughout his tenure at the Museum, Brown never lost his sense of adventure or failed to eagerly anticipate the prospect of a new locale, a new geologic horizon, or a new fossil find. His excitement comes through in an 1897 letter he wrote to Osborn from Wyoming: "Wish you were here to enjoy the pleasure of taking out those beautiful black, perfect bones, also to eat some of the juicy antelope stake [sic] I got Sunday." And his sentiments are still unmistakable thirty-six years later in a letter he wrote as he rode the rails west from Minneapolis on the New North Coast Limited toward Montana. "I am feeling fine and raring to get at the dinosaurs again," the fifty-nine-year-old Brown announced, adding, "as usual I am up with the dawn when nosing toward the fossils."

COMO BLUFF: A JURASSIC MENAGERIE

If Brown hadn't already decided he wanted to be a paleontologist, his first expedition as a full-time Museum staff member would have convinced him. In the summer of 1897, Osborn sent Brown to Como Bluff, Wyoming, home to the Late Jurassic fossil beds that had been the site of some of Othniel Charles Marsh's most spectacular finds of sauropods (the heavy-bodied, long-necked group of dinosaurs to which *Apatosaurus* belongs) and other dinosaurs.

At first Brown had little success and began to wonder if perhaps Marsh's teams had exhausted the site. But then, in a June 14 letter to

ABOVE: Museum explorers Jacob Wortman, Walter Granger, and Peter Kaisen unearthing *Apatosaurus* from Bone Cabin Quarry, Wyoming, one of the richest fossil sites on earth.   *LEFT:* The Museum's remarkable *Apatosaurus* (once called *Brontosaurus*), shown here with the wrong head—an error not corrected until the 1990s.

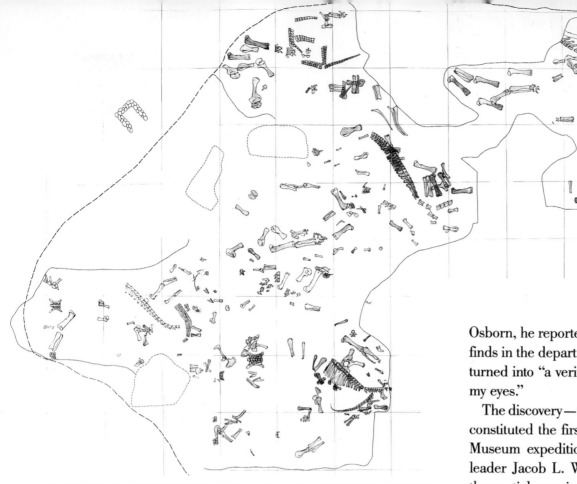

ABOVE: The abundant bones of Bone Cabin Quarry. The cabin itself, built out of fossil bone fragments by a local shepherd, is illustrated in the upper left of this sketch.

ABOVE: Skull of the great meat-eater *Allosaurus*, one of the dinosaurs found in the Bone Cabin trove.

Osborn, he reported what must stand as one of the most significant finds in the department's history. The quarry, Brown reported, had turned into "a veritable gold mine and I have been in bones up to my eyes."

The discovery—thirty feet of vertebrae belonging to a sauropod—constituted the first important dinosaur specimen unearthed by a Museum expedition. As the summer progressed, Brown, group leader Jacob L. Wortman, Walter Granger, and others excavated the partial remains of two enormous sauropods from the quarry. Osborn, knowing that the Museum would finally have its own dinosaurs to display, even made a rare journey out to visit the site.

The Como Bluff area proved to be one of the most productive areas ever excavated by Museum paleontologists. In 1898, prospecting north of the bluff, they came upon a hilly area so rich in fossils that dinosaur bones were simply strewn across the ground. Nearby stood a shepherd's cabin built entirely of fragments of dinosaur fossils, the structure that gave the famous Bone Cabin Quarry its name.

For seven years, Museum parties excavated the Bone Cabin site and surrounding areas. In 1898 alone they collected 141 specimens and sent one hundred boxes of bone and rock weighing a total of sixty thousand pounds back to New York City aboard J. P. Morgan's special railroad cars. By the time the quarries finally petered out, the Museum had acquired an extraordinary collection of North American Jurassic dinosaurs, including the great sauropods *Apatosaurus* and *Diplodocus*, the fleet predator *Ornitholestes*, and the powerful carnosaur *Allosaurus*.

LEFT: Allosaurus feasting on remains of an *Apatosaurus* in one of the most lifelike displays found in the Museum's old halls. It was an accurate display, too: Scientists have found *Apatosaurus* bones with *Allosaurus* toothmarks in them. *BELOW AND BOTTOM LEFT:* One of the Museum's most important early discoveries, *Ornitholestes* was a sleek and agile six-foot-long dinosaur that, while not turning down dead meat, may also have chased and captured smaller prey with its long-fingered grasping hands.

THE WANDERER

Barnum Brown himself didn't spend much time at Como Bluff after 1897. Characteristically, "Mr. Bones" was dreaming of new horizons even as important fossils were being excavated. In August 1897, just two months after the first Como Bluff find, he was already writing to Osborn, "Dr. Wortman informs me that you are thinking strongly of sending someone to South Africa, and I told him to write you that I wanted very much to go."

The Museum didn't send Brown to South Africa, but it did provide him with an equally exciting opportunity. As Brown himself wrote in one of the several versions he told of the story, "Arriving at the museum one morning, at 9 o'clock, Professor Osborn called me into his office, saying, 'Brown, I want you to go to Patagonia with the Princeton expedition. You may be gone a year and a half. The boat sails at 11. Can you make it?'"

Brown, of course, made it, and by the time he returned to New York City in 1900, he'd collected fossils of many ancient Patagonian mammals, notably the glyptodont *Propalaehoplophorus* (a mammal with armored plates, like a modern armadillo) and the rabbitlike

Pachyrukhos. Even more importantly, Brown's experiences in Patagonia irrefutably established his adaptability, resilience, and love of adventure. These were qualities that Osborn recognized and valued, and Brown put them all to use when he returned to North America in 1900 and began an extraordinarily productive twenty years of fieldwork.

His first important destination was the Hell Creek region, north of Miles City, Montana, in 1902. On a scouting trip in the Miles City area, Brown had identified exposures of rock in the Missouri Breaks dating from Late Cretaceous times, which he thought might contain dinosaurs far different from those excavated from the Jurassic beds of Como Bluff. Intriguingly, ranchers from these remote badlands had reported finding dinosaur bones, although details of such finds were sketchy.

After a strenuous week's travel by horse-drawn wagon from Miles City, Brown reached the Hell Creek region. Almost immediately he began to unearth fragmentary remains of the horned dinosaur *Triceratops*, but these exciting finds were soon cast into shadow by one of the greatest discoveries of his career: the first specimen ever found of the magnificent carnivore *Tyrannosaurus rex*.

Dressed exquisitely as usual, Barnum Brown surveys a promising fossil in Montana. His fine clothes and studious demeanor, however, masked an unmatchable drive to find and excavate the bones of ancient animals.

RIGHT AND ABOVE: *Triceratops* skull, and the living animal as imagined by Charles R. Knight. Brown and other paleontologists exploring the Late Cretaceous fossil sites of North America frequently found bones of this abundant dinosaur, but skulls were far less common, and perfect skulls like this one were more precious still.

THIS PAGE, COUNTERCLOCKWISE FROM TOP RIGHT: Whether in situ, getting their photographs taken, or in artists' fanciful restorations, Brown's two *T. rex* skeletons (one of which now resides at the Carnegie Museum in Pittsburgh) are etched indelibly in the imaginations of millions of dinosaur enthusiasts.

It took Brown almost two seasons of fieldwork to excavate the *Tyrannosaurus* from the nearly impenetrable sandstone that covered it. Using dynamite, he was finally able to free the bones and prepare them for shipping. The pelvis, however, weighed in at two tons and was too heavy to load aboard the wagon. Instead, the resourceful Brown built a sled and had a team of horses drag the fossilized bone to Miles City, which was more than one hundred miles away. The procession eventually reached its destination, where the skeleton was loaded aboard a train for shipment back east.

In 1908, Brown returned to the Hell Creek exposures and discovered another, even more perfectly preserved *Tyrannosaurus*—his "favorite child," as he referred to it in later years—which is the one on display in the Museum's halls.

While Brown spent much of the 1910s and 1920s prospecting in Wyoming, Montana, and elsewhere in western North America, he did venture abroad several times. One brief trip to Cuba in 1911 provided more proof of the paleontologist's indefatigable nature.

On one of his rare hunting trips for mammal—not dinosaur—fossils, Brown chose to explore a series of springs called Baños de Ciego Montero, which he translated as "Bath of the Blind Field Man." Here he faced a knotty problem: how to find and remove the bones from a site perpetually flooded with more than six feet of water.

Over the course of the next few weeks, Brown and his team wrestled with balky pumps, sudden rainstorms, and flooding from a nearby river. "Do not think I am having a vacation on this piece of work for we are at the pumps part of the night, as well as all the day and it is still just an even break as to whether the water floods us out," he grumbled in a letter to Osborn. In reply, Osborn, with typical sensitivity to the moods of the Museum's premier fossil hunter, thanked Brown for his "heroic efforts in overcoming the difficulties of the situation."

Brown, of course, did overcome these difficulties, and by the end of the damp dig, he had accumulated a fine collection of Pleistocene animals, including ground sloths, rodents, birds, and alligators. One important specimen, the sloth *Megalocnus,* is now on display in the Museum's new early mammal hall.

ABOVE AND BELOW: Brown did whatever it took—including rigging a pulley system to haul out *T. rex*'s pelvis or devising a pump system to drain a Cuban spring—to collect fossils. As Henry Fairfield Osborn put it in a letter to "Mr. Bones": "I have observed about you that difficulties appeal to you only as something to be triumphed over."

ABOVE: *Corythosaurus*, one of the bizarre hadrosaurs (duck-billed dinosaurs) that are found in such abundance in the Cretaceous exposures along Alberta's Red Deer River.

BROWN'S GLORY: THE RED DEER RIVER

Although Como Bluff provided Brown with his first dinosaur fossils and Hell Creek yielded his most famous discovery, *Tyrannosaurus*, no site meant more to Brown or produced a more spectacular haul of dinosaur bones than the extraordinary Cretaceous beds he prospected in the Red Deer River region of Alberta, Canada, between the years 1910 and 1915.

As so often happened, the discovery of the Red Deer River site was a stroke of good fortune. Visiting the Museum in 1909, a Canadian rancher from the Red Deer River valley had remarked that he had found many bones on his ranch quite similar to those on display in the Museum's dinosaur halls. This casual observation sparked Brown's interest—so much so that he scheduled a visit to the area in 1910.

The trip, however, proved a bit more difficult than he had anticipated. The Red Deer River cuts a deep, almost impenetrable valley through the plains of southern Alberta, creating a sheer drop of five hundred feet in some places and clearly not permitting exploration by horse or wagon. Brown, however, did not allow such obstacles to stop him from finding fossils. Since he couldn't get there by foot or on horseback, he traveled by boat.

Assessing the situation, the innovative paleontologist designed an unwieldy but perfectly adapted vessel for navigating the river: a twelve-by-thirty-foot flatboat equipped with a twenty-two-foot-long oar on either end and a sheet-iron stove in the middle. Setting off from the town of Red Deer, the Museum team left civilization behind, floating, as Brown recalled, "through picturesque solitude unbroken save by the roar of the rapids."

The method of exploring the area proved to be not only enjoyable but also extremely productive. The banks and canyon walls of the Red Deer River provided some of the richest exposures ever found of Cretaceous fossils. In two seasons of fossil hunting by boat, Brown brought home many tons of dinosaur bones, particularly those of the hadrosaurs, or duck-billed dinosaurs, which had clearly been abundant in the region.

LEFT: The Red Deer River sites were inaccessible by land, so Brown built a boat—not the world's most streamlined vessel, but just what the collector needed to reach some of the richest fossil sites in North America.

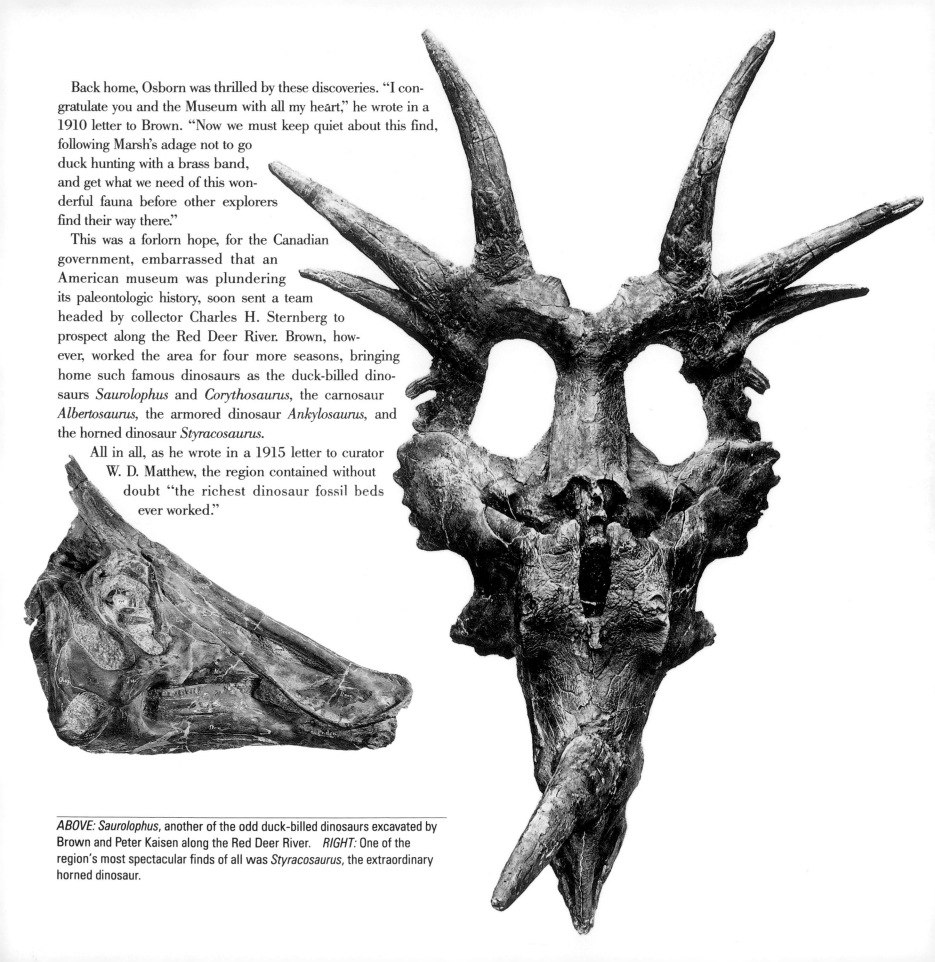

Back home, Osborn was thrilled by these discoveries. "I congratulate you and the Museum with all my heart," he wrote in a 1910 letter to Brown. "Now we must keep quiet about this find, following Marsh's adage not to go duck hunting with a brass band, and get what we need of this wonderful fauna before other explorers find their way there."

This was a forlorn hope, for the Canadian government, embarrassed that an American museum was plundering its paleontologic history, soon sent a team headed by collector Charles H. Sternberg to prospect along the Red Deer River. Brown, however, worked the area for four more seasons, bringing home such famous dinosaurs as the duck-billed dinosaurs *Saurolophus* and *Corythosaurus*, the carnosaur *Albertosaurus*, the armored dinosaur *Ankylosaurus*, and the horned dinosaur *Styracosaurus*.

All in all, as he wrote in a 1915 letter to curator W. D. Matthew, the region contained without doubt "the richest dinosaur fossil beds ever worked."

ABOVE: Saurolophus, another of the odd duck-billed dinosaurs excavated by Brown and Peter Kaisen along the Red Deer River. *RIGHT:* One of the region's most spectacular finds of all was *Styracosaurus*, the extraordinary horned dinosaur.

THE VALUE OF A DOLLAR

Barnum Brown's accomplishments are especially remarkable given the meager amount of money earmarked for his early expeditions.

In an unsigned letter dated June 29, 1900, the Department of Vertebrate Paleontology outlined a budget for Brown's three-month expedition to the Laramie Formation in South Dakota, scheduled for July 1 through October 1. The budget stated that Brown would be allowed to spend no more than eighty dollars on provisions, two hundred dollars for "Help," and forty dollars for saddles. "The expedition including everything is not to exceed $800, and to be kept as far within this limit as practicable," the letter dictated.

Months later, Brown submitted his expenses for the Laramie expedition. The following were among the costs he charged to the Museum:

Pick, shovel, and hammer	*$ 4.25*
Hotel	*$ 2.00*
Cook "Armstrong"	*$18.00*
Blasting powder and fuse	*$ 1.00*
Cement and oats	*$ 4.00*

His total expenses came to $801.60, or $1.60 more than the Museum had allotted him. Clearly, Barnum Brown was not able to keep to a budget. Upon closer inspection, however, Brown's Laramie trip lasted from July 1, 1900, to January 26, 1901—nearly seven full months. For a mere $1.60, the young paleontologist had tacked on four additional months of work, bringing back bones belonging to *Triceratops,* hadrosaurs, and other important dinosaurs.

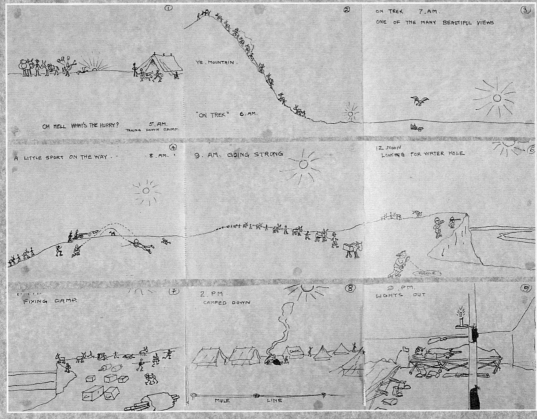

A typical day in the field for Brown and his intrepid band of adventurers.

By the 1930s, Barnum Brown (shown above, at right, in the laboratory) was the most famous fossil hunter on earth, his exploits making the front page of major metropolitan newspapers.

THE WORLD'S MOST FAMOUS FOSSIL HUNTER

Brown's fame continued to grow after his years on the Red Deer River. He traveled and prospected as actively as ever, journeying to India, Greece, and other countries in the 1920s and excavating a magnificent trove of Jurassic fossils from Howe Ranch, Wyoming, in the 1930s, but his most celebrated discoveries were behind him.

Partly, this was due to circumstances he could not control. The Museum sharply curtailed its collection budget after the onset of the Great Depression, and with the notable exception of a series of productive expeditions cosponsored by the Sinclair Oil Company, outside sponsorship was also hard to find.

Brown's focus also shifted as he grew older, however. He seemed to become more enamored of the publicity his expeditions created, perhaps at the expense of the finds themselves. In the late 1920s, for example, he spent months traveling around the United States, but his destinations were town halls and lecture sites, not new fossil horizons.

Brown retired from the Museum in 1942, but for years afterward, he hoped for funding for one more major expedition. Even without financial support, the seventy-nine-year-old Brown still managed to collect and send a load of Pleistocene mammal fossils back to the Museum from Guatemala in 1953.

Barnum Brown died just shy of his ninetieth birthday. At the time of his death, it was said, he was planning a fossil-hunting expedition to the Isle of Wight. Brown was certain he would find dinosaur bones there.

OTHER EXPLORERS, OTHER HORIZONS

These specimens that cost me so much in labor, exposure and untold hardships never hurt my sensitive heart. The glory of looking for the first time on many of them compensated me for all my troubles, and I love them yet.

Charles H. Sternberg, 1902

Barnum Brown's extraordinary talents—as well as his eye for publicity—make it easy to forget that he was not the only paleontologist collecting dinosaur fossils for the Museum during the vertebrate paleontology department's glory years of 1890 to 1930. Each year, as Brown set out for Wyoming, Montana, Alberta, or more far-flung destinations, department curators such as W. D. Matthew, Walter Granger, Bill Thomson, and even Henry Fairfield Osborn himself prospected in fossil beds in Texas, Kansas, New Mexico, and many other states. Most of the finest mammal fossils in the Museum's collection—and no small number of the dinosaurs—were brought home from these expeditions.

Remnants of an earlier age: diaries, field journals, and equipment from the great era of paleontological discovery.

Wasatch

Torrejon

Puerco

Record of Specimens

New Mexico
1913

Party

Walter Granger
George Olsen
Wm J. Sinclair
John Martin

H. F. Osborn
H. O. ...

39 Rodent skull frag. South Elk Cr. July 17
40 16926 gigantic bird skull part skul. 7 pkts. South Elk Cr. July 20
41 16958 Carnivora Skull fra South Elk Cr. July
42 small jaws South Elk Cr. July
43 16975

46 16967 Carnivora jaw frag South Elk Cr. July 25
47 small jaws teeth frag South Elk Cr. July 26
Coryphodon lower jaws South Elk Cr. July 26
Oxyaena Carnivora lower jaws Elk Cr. July 27
teeth frag Cr. July 28
July 29

...mie Basin 1898

...t. Hist
...al Park W...
New York

ABOVE: Charles H. Sternberg was moody and difficult, but possessed an unquenchable tenacity and devotion to Henry Fairfield Osborn. "I candidly tell you that I would rather collect fossils for the American Museum than any institution on earth," he wrote to Osborn.

RIGHT: "The boy George of mine whose eyes I have trained for many years," said Charles H. Sternberg of his son. The training proved worthwhile, as George became a prominent fossil hunter in his own right.

In addition, not all of the most important fossils were collected by members of the department staff. Of course, Osborn would have preferred to rely entirely on his own collectors, because all of their finds automatically belonged to the Museum. Sometimes, however, free-lance fossil hunters, ranging from geologists on state survey teams to full-time paleontologists who collected for many different museums, made discoveries that Osborn wanted to obtain for the Museum's collection.

CHARLES H. STERNBERG'S ENDLESS QUEST

The names of many of these free-lance collectors are unfamiliar today, but one, Charles H. Sternberg (1850–1943), is still remembered by dint of his longevity, his vivid personality, and his nearly unparalleled ability to locate valuable fossils. Sternberg was responsible for finding many of the Museum's prize specimens, including what he called the "crowning specimen" of his life's work: the "mummified" *Edmontosaurus* (most recently known as *Anatosaurus* and originally called *Trachodon*).

Like Barnum Brown, Sternberg grew up in Kansas, collected fossil shells as a boy, and was determined from an early age to be a professional fossil hunter. After graduating from college in 1871, he joined Edward Drinker Cope's collecting team on an expedition in Kansas and spent the rest of his long life (often accompanied by his three paleontologist sons) digging up bones in Texas, Alberta, and many other locations. Today, his dinosaurs and other ancient creatures populate more than two dozen museums in the United States and Europe.

At first glance, Sternberg appeared to be a man who had achieved all his goals, but from the books, articles, and letters he wrote during his career, it is clear that he was a troubled individual whose life was filled with contradictions. Enamored of the thrill of fossil hunting, Sternberg said he never regretted the career choices he had made, but he also felt that his work was never fully appreciated. He was a deeply religious man whose writings are filled with acknowledgment of the powers of God and belief in Creation, yet he devoted his life to unearthing irrefutable evidence of evolution. He idolized Henry Fairfield Osborn and the Museum, but spent years railing against the Museum's inability to pay what he thought he deserved.

ABOVE, LEFT AND RIGHT: A magnificent *Triceratops* skull, one of Charles H. Sternberg's hundreds of fine discoveries over a decades-long career. *LEFT:* Sternberg and sons in a typically spartan camp in Alberta. Note that the team was working under two flags: the Stars and Stripes and the Union Jack.

Sternberg's constant money woes left a lasting mark on his personality. Barnum Brown, W. D. Matthew, and other curators might have complained about their salaries, but they were secure in the knowledge that they received steady wages for doing what they loved. In contrast, Sternberg, who would not receive a penny if he failed to sell any of his finds at the end of a costly field season, was legitimately poor. "He is penniless and completely worn out," reported a Museum curator after visiting the seventy-three-year-old collector's camp in New Mexico.

Throughout his life, Sternberg was constantly aware of his plight—and his disadvantages compared to full-time Museum staff. In nearly every letter he wrote to Osborn over a thirty-year period, he mentions his poverty, a constant rehashing that sounds like a mix of recrimination and self-flagellation. In fact, many of Sternberg's letters are so naked in their emotions that they are difficult to read, even today. One can imagine that Osborn dreaded opening them.

If that is true, however, Osborn never let Sternberg know it. Part of what made Osborn such a brilliant administrator was his ability to answer his collectors' complaints with politeness, restraint, and unending support. His letters to Sternberg are invariably kind and sympathetic, and it is clear that on occasion he purchased one of the collector's finds more out of loyalty than out of actual need for the specimen.

Sternberg's eccentric collecting habits, however, did not make Osborn's job any easier. He tended to immediately pack his specimens in plaster and wooden crates and then ask large amounts of money for them, sight unseen. In addition, he reveled in his discoveries so much that he tended to overvalue them, which made negotiations with the Museum painful and protracted.

In the end, Osborn's patience with the collector was rewarded, and throughout his life, Sternberg's loyalty to the Museum remained unquestionable. His expeditions to the Red Deer River region in Alberta, to the Permian fossil beds of Texas, to New Mexico, and to sundry other locales have helped populate the Museum's halls with *Albertosaurus*, *Pentaceratops*, and the great *Edmontosaurus* "mummy," among many other specimens.

Sternberg lived to see his finds achieve the renown they deserved. When he was eighty-eight, he was invited by Barnum Brown to visit the redesigned vertebrate paleontology halls. Declining, he wrote, "How I would like to see the new arrangements of the Dinosaurs. But my greatest pleasure in looking back on my life is to know that some of my discoveries sit preserved in the American Museum."

LIFE DURING WARTIME

The century-long history of the Department of Vertebrate Paleontology has spanned two world wars, yet one would scarcely know it from reading the field notes, correspondence, and published papers of the department's curators. While making allowances for shortages of fuel and other supplies and occasionally being forced to change their travel plans, Barnum Brown, W. D. Matthew, and others simply went about their business, barely acknowledging the great struggles that dominated every news report.

In fact, of all the hundreds of letters sent to and from the Museum during wartime, only one uncharacteristically gloomy assessment by Matthew confronts the issue directly. "The war is evidently going to be a prolonged and exhausting struggle, and the longer it lasts the harder we shall be hit by it," he wrote to Brown in 1914. "Science in Europe will be as dead as it was in the Southern states for nearly half a century after the Civil War."

Perhaps the First World War's most dramatic effect on the Department of Vertebrate Paleontology actually took place in 1920, when the department received an urgent, almost panicky letter from the great Belgian paleontologist Louis Dollo. "I learn that the situation is *terrible* in Vienna and that our friend *Abel* is in danger!" Dollo wrote. "We cannot leave him to perish."

Othenio Abel, a good friend of Henry Fairfield Osborn's, was one of the leading European paleontologists of his time, and his theories had influenced the work of many of the Museum's curators. Leaping into the breach, the curators contributed a total of three hundred dollars (with Osborn himself contributing one hundred dollars), which purchased hundreds of pounds of flour, beans, bacon, milk, and other food supplies for Abel and his family.

Abel was stunned by the gift, which he had not expected. "Your people have done a great deal for our children," the translation of his grateful letter reads, "and in this way have aided materially in reducing the number of cases of sickness due to privation and hunger."

OPPOSITE, RIGHT: The evolution of the horse, from *Hyracotherium* to *Equus*—a gradual increase in size, decrease in number of toes, and changes in tooth shapes, all used by Museum paleontologists as conclusive proof of evolution. *OPPOSITE, LEFT: Homo erectus* hunters pursuing *Hipparion*, an ancient relative of modern horses.

HORSES AND THE BATTLE OVER EVOLUTION

Henry Fairfield Osborn sought few fossils more enthusiastically than those of ancient horses. Early in his reign as curator of the Department of Vertebrate Paleontology, he dreamed of large-scale collection and restoration efforts centered on horses, entire displays in new halls devoted to horses, and books on horses published under the auspices of the Museum. One of his first fund-raising efforts was an unsuccessful appeal to the trustees for the grand sum of ten thousand dollars—for a project on horses.

From a modern perspective, this focus on a single family of mammals seems a bit narrow. On second look, however, Osborn's love of ancient horses is entirely understandable. Early on as a student of paleontology, he had been fascinated by the ongoing debate over Darwin's theories of evolution, and no line of animals seemed to provide clearer evidence of evolution than horses.

Today's scientists paint a far more complex picture of horse evolution than paleontologists did a century ago, but, at least broadly, the evolutionary trends that so excited Osborn can still be seen.

The horse line begins with the primitive, terrier-size *Hyracotherium* (previously called *Eohippus*, the "dawn horse"), which had three toes on its hindfeet and four on its forefeet, and low-crowned, squarish molars. As the family evolved, from *Orohippus* through

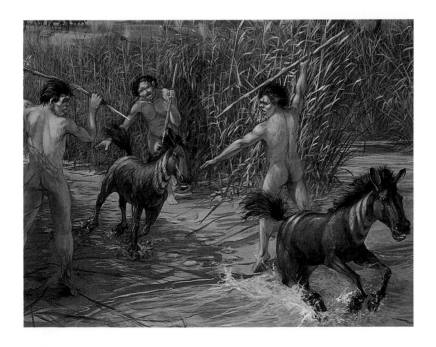

Mesohippus, *Parahippus*, *Protohippus*, *Dinohippus*, and *Equus*, fossils reveal a gradual tendency toward greater size and height, increase in complexity of the molars (adaptations for more efficient grazing on tough grasses), and reduction in number of toes to the single-toed hoof found in modern horses.

To make this evidence of evolution clear to Museum visitors, Osborn succeeded in wangling fifteen thousand dollars from trustee William C. Whitney in 1897. He used the money to send newly hired collector James W. Gidley to Texas and South Dakota, while curators Bill Thomson, W. D. Matthew, and Walter Granger were constantly on the lookout for fossil horses during their collecting expeditions to Colorado, South Dakota, and elsewhere.

As a result of these intensive efforts, the Museum's collection of fossil horses grew exponentially; by the early 1900s, the collection included dozens of specimens of many individual genera, including *Hyracotherium*, *Pliohippus*, *Hypohippus*, *Neohipparion*, and *Equus*.

Osborn had many of the skeletons mounted in the most lifelike postures possible and missed no opportunity to promote both the displays themselves and their importance as examples of the evolutionary process. He and other department curators wrote on the subject for scientific journals and the popular press and gave lectures to universities and horse clubs alike in the first decade of this century. In the end, the Museum's horse displays were among the best-known exhibitions in the world.

Unsurprisingly, despite Osborn's most diligent efforts, the debate over evolution was not put to rest by the Museum's fossil horse exhibits. Rather, the conflict continued to run like a steady stream through the history of the Museum, with curators having to return to the subject again and again in response to public questions, complaints, and accusations.

Many of the doubters were polite, questioning specific details rather than the theory as a whole. "I am not yet satisfied that all of evolution can be accounted for by external stimuli," wrote former United States and Museum president Theodore Roosevelt to W. D. Matthew in 1915.

TOP AND ABOVE: W. D. Matthew and George Gaylord Simpson: two of the Museum's finest paleontologists, as well as its most ardent defenders of evolutionary theory.

The Museum, however, was also a constant target of those individuals and groups who viewed belief in evolution as sacrilegious. Matthew, for one, had no patience with these doubters, refuting them in articles, book chapters, letters, and even satirical verse. In a mocking poem written from the point of view of a panicked anti-evolutionist, Matthew wrote, "And next the scientist will find/The evolution of our kind/Goes back to arthropod or worm/That in the Cambrian slime would squirm/Oh come! Let's stop it at the ape/At least he's got some human shape."

More seriously, others argued that it was possible to believe in both evolution and God. As early as 1902, Charles H. Sternberg, an extremely religious man, referred to his fossil finds as the "wonderfully preserved records of the Almighty in the rocks of America."

But perhaps the most eloquent defender of what might be called the "Unified Theory of Religion and Evolution" was Museum curator George Gaylord Simpson, who returned often to the subject during his long career. In a 1946 response to a letter published in *Natural History* magazine, for example, Simpson wrote, "Some scientists are agnostics or atheists, just as some business men or some mechanics are agnostics or atheists ....I do know that many outstanding scientists believe in God."

Today, according to curator and Department of Vertebrate Paleontology chairman Richard Tedford, the evolution debate doesn't come up as often as it once did. "We hardly ever get any letters on the subject any more," he says. "Perhaps the issue is finally close to being settled—or perhaps those who disagree with us simply have given up hope of changing our minds."

CHILDS FRICK

A talented paleontologist, Childs Frick used his great personal wealth to sponsor some of the twentieth century's most important scientific expeditions in search of ancient mammals in China, Alaska, and across the lower forty-eight states. Further, without his and his corporation's generosity, the Department of Vertebrate Paleontology's home at the American Museum of Natural History—the Frick Building—would not exist.

Against the wishes of his father, financier Henry Clay Frick, Childs Frick chose to pursue his love of science after his graduation from Princeton in 1905. At first, he concentrated on living mammals, collecting specimens for the American Museum and other institutions in East Africa and elsewhere.

But soon Frick succumbed to the siren song of the past and changed his focus to extinct mammals, particularly those dating from the Late Cenozoic period. By the 1930s, while sponsoring his own team's expeditions under the auspices of the Frick Laboratory, he had established or supported more than two dozen research funds at the Museum. These provided support for a variety of important expeditions, including Roy Chapman Andrews' Asiatic forays, but the most important was the extensive American Pliocene Fund, designed to sponsor Late Cenozoic research.

Frick's own collecting teams were working year-round by the mid-1930s. In the winter months, they would unearth fossils in California, Arizona, Texas, and Florida, and then switch to Nebraska, Wyoming, New Mexico, Montana, and other states in the summer.

The sheer quantity of important fossils collected during a single field season was often extraordinary. According to the *Farthest-North Collegian*, published by the University of Alaska, for example, the 1938 Alaska field season shipped a massive eight tons of Pleistocene fossils back to the Museum. "Besides the usual mammoth, super-bison, caribou, and horse fossils," the newspaper reported, "this year's collection includes such rarer finds as the mastodon, musk-ox, elk, moose, bear, lion, sabre-tooth tiger, camel, wolf, fox, wildcat (probably lynx), ground squirrels, and field mice."

Childs Frick was a scientist, a philanthropist, and a lifelong fossil enthusiast, particularly of ancient mammals of North America.

By the time of Frick's death in 1965, the Frick Collection contained as many as a half million individual specimens—and it is unlikely that they will all ever be counted to obtain a more precise number. Following Frick's wishes, this entire collection was given to the Museum in 1968, and soon thereafter, the assets of the Childs Frick Corporation were transferred to the Museum to endow the collection.

Today, the influence of Childs Frick on the Department of Vertebrate Paleontology is immediately apparent to anyone who wanders the halls of the Frick Building. Open one of the storeroom cabinets or peek inside one of the rough-hewn crates that line the walls of the building's fourth floor, and you'll gaze on some of the countless fossils that are his lasting testament.

AGATE SPRINGS: CRATES FULL OF GEMS

Of all the fossil troves yet found in North America—including Wyoming's Como Bluff, Montana's Hell Creek, and California's La Brea tar pits—one of the richest and certainly one of the least well known is located in Sioux County, Nebraska.

In the past century, the Miocene fossil beds near Agate Springs have provided museums with perfectly preserved skeletons of nearly a dozen different Miocene mammals, ranging from early rhinoceroses to primitive horses, beavers, and camels. Most of these have been found in such great numbers that they allow us to envision a vivid picture of mammalian life in midwestern North America twenty million years ago.

Unlike such sites as Como Bluff, which the Museum alone prospected in the late 1890s, or the Red Deer River area in Alberta, which Barnum Brown had to himself for two years, exploration of Agate Springs was well under way by the time the Museum's Bill Thomson and W. D. Matthew arrived there in 1908. Fossils were first discovered in the area in 1877, and by the early 1900s, paleontologists from the University of Nebraska, the Carnegie Museum in Pittsburgh, Yale University, and Amherst College were all excavating quarries within a bone's throw of each other.

Sharing a site was an unpleasant new experience for the Museum collectors. Matthew's 1908 letters to Henry Fairfield Osborn are filled with barely masked annoyance, as if the curator felt strait-jacketed by the presence of such contemporary paleontological luminaries as Yale's Professor R. S. Lull, Nebraska's Professor E. H. Barbour, and Amherst's Professor F. B. Loomis.

Matthew was annoyed enough by the situation in 1908 to suggest that the Museum establish a permanent museum at the Agate Springs site. Matthew foresaw many advantages to the plan: "If run by the A.M.N.H. it would be a great advertisement to us as well as a fine educational influence." Then, however, came the kicker: "If we don't do it, someone else probably will."

Osborn was favorably disposed to the idea, but nothing ever came of it. Still, despite Matthew's frustration, there were more than enough bones to go around. In 1908, Matthew and Thomson sent home nine skeletons of the camel *Stenomylus* from a rich deposit a few miles from Agate Springs. Today, this group of skeletons is one of the highlights of the newly redesigned mammal halls.

ABOVE: The extraordinary quarry of bones of *Diceratherium*, the small rhinoceros that populated the plains of what is now the Agate Springs area in Nebraska.

Thomson returned many times to Agate Springs in the summers following his initial visit in 1908. After nearly twenty years of field-work in the area, he had collected many fine specimens for the Museum, including the bizarre-looking plant-eating *Moropus* (a distant relative of modern horses), and a slab of rock containing the skulls and other bones of at least twenty-two *Diceratherium*, a small two-horned rhinoceros. A free-mounted skeleton of *Moropus* and the *Diceratherium* slab can be seen in the Museum's halls today.

BELOW: An Agate Springs scene, courtesy of Charles Knight, with the camel *Stenomylus*, the rhinoceros *Diceratherium*, the horse *Merychippus*, the pig *Dinohyus*, and *Moropus*, a relative of modern horses.

HOW TO BE A HEALTHY PALEONTOLOGIST

In a 1922 letter to ailing research assistant Charles C. Mook, Henry Fairfield Osborn dictated his rules for maintaining good health and fitness.

(1) Rise an hour before breakfast and take a half hour of light walking and breathing exercise, dressing in good warm flannels so as to induce perspiration.

(2) Return home, take a tepid bath and a good hard rub down.

(3) Eat slowly a hearty breakfast—cereals, milk, cooked fruit only, brown or whole wheat bread only. A half hour's amusement afterwards.

(4) Absolutely give up the use of coffee in any form or at any time of the day. Avoid smoking and other stimulants.

(5) Follow your breakfast by three hours of garden or farm exercises; no newspapers, correspondence or literary work, unless absolutely necessary.

(6) A hearty noon dinner at 1 o'clock, the principal meal of the day, followed by a half hour of leisurely light reading or conversation—no work.

(7) Afternoon as you please until 5:00 P.M. Supper at 6:30 P.M., plenty of milk and vitamins; cooked fruits only; no tea or coffee.

(8) Light reading and family amusement only during the evening; cards or checkers; absolutely no literary or scientific work of any kind—nothing serious.

(9) To bed by half past nine o'clock.

There is no record as to whether Charles Mook followed Osborn's advice. Mook did, however, remain at the Museum in apparent good health until 1938 and was still producing papers in the 1960s, so presumably he was doing something right.

ROY CHAPMAN ANDREWS: THE LAST GREAT ADVENTURER

*I was born to be an explorer.
There was never any decision to make. I couldn't do
anything else and be happy.*

Roy Chapman Andrews, in
This Business of Exploring, 1935

As the 1910s drew to a close, so did the great era of dinosaur hunting in North America. Red Deer River, Hell Creek, Como Bluff, and other famous locations were tapped out, and with few exceptions, new and equally rich sites had not been found. Barnum Brown and the other department curators were still prospecting, still bringing home fossils, but their work had lost the pioneering aura that had enveloped it just a few years before.

A camel caravan bearing supplies for Roy Chapman Andrews' Central Asiatic Expeditions wends it way through the Gobi Desert of Mongolia.

THIS PAGE: Roy Chapman Andrews (top, at right, and above left), in *This Business of Exploring*: "I have been so thirsty that my tongue swelled out of my mouth....I have seen my whole camp swept from the face of the desert like a dry leaf by a whirling sandstorm. I have fought with Chinese bandits. But these things are all a part of the day's work." Doesn't that sound a bit like Indiana Jones (above right)?

The lull, however, did not last long. Enter Roy Chapman Andrews and his extraordinary series of expeditions to the Gobi Desert of Mongolia between 1919 and 1930. Without doubt, these were the most exciting—and the most widely publicized—adventures ever sponsored by the Museum. Children worldwide idolized Andrews and dreamed of following in his footsteps, of crossing a great untracked desert by motorcar and camel caravan, of fighting off marauding bandits, of discovering fossils no one else had ever seen.

To be sure, Roy Chapman Andrews cut a figure that was larger than life. It is even claimed that this explorer extraordinaire was the real-life model for the movie character Indiana Jones. If he wasn't, he should have been.

Andrews said that his desire to be an explorer dated from his childhood in Beloit, Wisconsin, and that it was his lifelong ambition to work at the American Museum. He moved to New York City in 1906 soon after graduating from college and took a job at the Museum as a custodian in the taxidermy department.

This was where he pursued his first passion: whales. During the next decade, he traveled all over the world aboard whaling ships to such places as the North Atlantic and the waters off Korea and Japan, and he wrote two of the very few scientific papers he would produce during his lifetime. Andrews was enthralled with the travel, the adventure associated with the study of whales; the science of mammalogy, which required painstaking work back home in the laboratory, was far less appealing.

Even as he explored these remote oceans, Andrews was making plans for an elaborate series of expeditions to the Gobi Desert of Mongolia. These expeditions, he hoped, would bring a squadron of scientists to the little-known desert, where they could study the region's geology, tally its bird and animal life, and perhaps even collect its fossils.

In 1919 and 1920, Andrews made a pair of reconnaissance and mammal-collecting trips into the desert. While on these journeys, he was captivated by the desert's beauty. "We trotted back to camp with the afterglow of the sun which painted the sky in streaks of crimson & gold," he wrote in his 1919 journal. "The night air was like a draught of wine after the heat of the day's sun." He also became convinced that a series of more ambitious expeditions was possible —and that he was the man to lead them.

Andrews' focus on the Gobi had originally been inspired by Henry Fairfield Osborn's belief that Asia, not Africa, was the birthplace of modern man, *Homo sapiens.* By 1920, Osborn was in the

THE FORMER
U.S.S.R.

MONGOLIA ○URGA
TSETSEWANG ○

SAIN NOIN KHAN ○ ○TUERIN

NOM MANCHURIA
KHONG ○IREN DABASU
SHABARAKH USU
CHINA
GOBI DESERT LATTIMORE'S ○ URTYN
LOST CITY OBO

○PEKING

------- ANDREWS
1922, 1923, 1925

ANDREWS
1928, 1930

INSET: The routes taken by the Asiatic Expeditions. *ABOVE:* Dwarfed by the eroded Red Mesa in the background, the explorers' tents provided little protection against fierce desert winds.

grip of the controversial racial theories that bedeviled him for the rest of his life, and he believed that the region Andrews hoped to visit would yield up the fossils of what the popular press called the missing link: the light-skinned, plains-dwelling humans who had been our true ancestors.

Osborn based his theories on no firm evidence: The Gobi desert and surrounding regions had produced no fossils of early hominids. In fact, the entire vast, unknown expanse had given up but a single fossil of any kind: a tooth of a primitive rhinoceros that dated from far before the earliest human.

With typical self-confidence, however, Osborn discounted the opinions of all who doubted him, including geologists who pronounced the Gobi an enormous pile of sand containing no fossils

at all. And the Museum's headstrong president was able to find a kindred spirit in Roy Chapman Andrews.

Even with Osborn's steadfast support, Andrews' ambitious plan faced a host of potential roadblocks. In 1920, Mongolia was represented as a huge blank space on most maps, and the arid area known as Outer Mongolia, which includes the vast Gobi Desert, was even less well known than the rest of the region. Inhabited by a sparse population of isolated nomadic peoples, the Gobi was rumored to be beset with armed bandits and renegade Chinese and Russian soldiers. If the bandits or soldiers didn't get you, the naysayers predicted, then the weather—frigid nights, broiling days, and sudden, deadly sandstorms—certainly would.

Andrews shrugged off such warnings and went about the task of raising money. A donation of fifty thousand dollars from J. P. Morgan, Museum funding, and fund-raising efforts turned his dreams into reality, and by 1922, Andrews had gathered his team of Museum scientists and staff and was ready to begin the first of five trips into the unknown.

The cast of researchers changed during the course of the Central Asiatic Expeditions. Among those whose names surface repeatedly are geologists Frederick K. Morris and Charles P. Berkey, photographer J. B. Shackelford, and—by all contemporary accounts, most centrally—vertebrate paleontologist Walter Granger.

Granger, a popular and well-liked scientist who was nearly fifty years old at the time of the 1922 expedition, had been at the Museum since 1890. Like Andrews, his first job was in the institution's taxidermy shop, where he was expected to shoulder any task requested by the senior staff members.

In a 1939 letter to a Museum publication called *The Grapevine*, he described one such task. Recalling that in the 1890s the walkway to a Museum entrance had been lighted by kerosene lamps on posts, Granger wrote that he was expected "to clean, trim and refill those lamps twice a week after open nights at the Museum—a nasty job during a bitterly cold winter and at the salary of $20 a month."

Moving to the Department of Vertebrate Paleontology in 1896, Granger soon became one of the Museum's most respected field researchers, overseeing expeditions to Wyoming, New Mexico, Utah, and elsewhere and helping to add to the Museum's growing collection of mammalian fossils. He also served as a teacher, mentor, and friend to many of the young paleontologists who would dominate the department in the years to come.

George Gaylord Simpson and Edwin H. Colbert were just two of the Museum scientists who remembered Granger fondly. In *Concession to the Improbable*, Simpson recalled, "Granger and I were soon Walter and George, and he became virtually a loved second father to me." And describing the Central Asiatic Expeditions in his *The Great Dinosaur Hunters and Their Discoveries*, Colbert added, "Granger by his very presence, by the force of his remarkable personality, helped immeasurably to make the work of the expedition go smoothly."

1922: SINK OR SWIM

On April 21, 1922, Andrews, Granger, and the rest of the Museum team headed through the gate of the Great Wall outside Beijing and onto the plains of Inner Mongolia. Although their immediate destination was the nearby town of Kalgan, where they were to pick up additional supplies, they knew they would soon be entering "a land of desolation, of thirst, and bitter cold and parching heat; of sandstorms and of tragedy," as Andrews described it in his usual dramatic prose.

Attempting to overcome these obstacles, the expedition party employed a brilliant combination of modern technology and time-honored desert tradition. Andrews had planned for the group to travel in a fleet of motor vehicles consisting of three Dodge cars and two Fulton trucks. Since there was no way these vehicles could carry sufficient supplies for the summer-long exploration, a caravan of seventy-five camels preceded the scientists, dropping food and gasoline at predetermined sites along the way. This plan allowed the explorers to penetrate farther into the Gobi than any previous group of scientists.

The Museum party didn't have far to go, however, before they realized that the expedition was, in fact, destined for the great success Andrews had prophesied. Only four days after setting out, the group camped at an abandoned oasis in a desolate basin known as Iren Dabasu, which was located 265 miles from Kalgan. That first evening, Granger, Morris, and Berkey did some preliminary reconnoitering of the area by truck, returning to camp at sunset.

In a story he told often, Andrews recalled that he could see at once that the three scientists were excited. As everyone watched, Granger reached into his pockets and pulled out a variety of fossils of ancient mammals. "Well, Roy, we've done it," he said. "The stuff is here."

Within a day, the party found several more fossils, including a bone belonging to a large dinosaur—the first dinosaur bone ever found in eastern Asia. None of the fossils belonged to the early humans Osborn had hoped (and expected) they would find in the Gobi, but to Granger and Andrews, these far older fossils were just as exciting.

A week later, before the last of the group left Iren Dabasu, they began to experience the harsh weather conditions that would threaten them throughout this and all succeeding expeditions. By that time, however, they already possessed a cheerful machismo

LEFT: Ehrlien, the last vestige of civilization, gateway to the empty expanses of Outer Mongolia.

RIGHT: Andrews (at center, standing) inspects the camel supply caravan, the group's lifeline in the months to come.

that allowed them to shrug off the severest weather conditions and other threats.

In a letter to the Museum from Iren Dabasu, Charles Berkey wrote, "Desert life is strenuous but we are all standing it….We have learned to eat sand with more or less relish in all kinds of food and to sleep peacefully with the chill winds simply sweeping through the tent and threatening to blow the flimsy looking, ballooning thing into the sea of Japan. Judging from the amount of wind that has come this way in the last week, there ought to be a great scarcity of atmosphere somewhere."

As the 1922 expedition wended its way into Outer Mongolia, the most dangerous threats to the party came from human interference rather than from the harsh physical conditions. Political intrigue, banditry, and conflicts between the Chinese and Russians were common.

Nowhere in their journals or letters do Andrews or any of his team seem worried by the dangers that lay before them. Back at the Museum, however, curator W. D. Matthew was not so sanguine. In an unusually stiff and formal letter to Walter Granger, Matthew wrote, "As you will recall, the understanding has been that you were to avoid risks….I want to say officially that your health and safety take precedence over any finds that you could hope to make, if only as a matter of keeping one of the most valuable members of our staff for the many future years that we hope to have him with us."

Granger's reply to this forlorn long-distance warning does not seem to have survived, but it probably will not come as too much of a surprise that the expedition members continued on without a single backward glance.

They continued on, in fact, to one of the most important sites visited by any of the five expeditions. In August, at a desert location dubbed Wild Ass Camp (so named because Andrews had spotted, and fruitlessly chased, three wild donkeys there), the party found one of the most spectacular specimens now on display in the mammal halls: bones of the gigantic ancient rhinoceros *Paraceratherium*, which they knew as *Baluchitherium*.

Such an important discovery rejuvenated the group. As Andrews wrote in his field journal on August 6, the day after they had excavated the animal's enormous skull: "At nine o'clock Shack, Walter, Bayard, Wang & I started—in the truck—for the *Baluchitherium* bed….We were a jolly party, Shack and Walter stretched out in camp chairs as we rolled along in the sunlight over the desert."

The discovery of *Paraceratherium* was the most dramatic find of the 1922 field season, but the most significant event of the summer took place in September, at the very end of the expedition, as the explorers headed back toward their headquarters in Beijing. The group made a brief stop in an area called Shabarakh Usu (now known as Bayn-Dzak), a site of almost mystical beauty. "From our tents, we looked down into a vast pink basin, studded with giant buttes like strange beasts, carved from sandstone," Andrews wrote. "There appear to be medieval castles with spires and turrets, brick-red in the evening light, colossal gateways, walls and ramparts."

Here, at the site the explorers named the Flaming Cliffs of Shabarakh Usu, photographer J. B. Shackelford wandered off and soon came upon a barren basin that could not be seen from the road. While there, he found a small white bone, which Walter Granger later determined was a portion of a skull of an unknown Cretaceous dinosaur.

At the risk of being trapped in the desert by autumn's severe weather, the expedition could not tarry long at the Flaming Cliffs. In the brief time they had at the site, however, the scientists found several more dinosaur bones as well as a piece of a fossil eggshell, which they all assumed belonged to some ancient bird. Taking this material with them, they headed onward, vowing to return.

Roy Chapman Andrews (at left) and Walter Granger inspecting bones of the great fossil rhinoceros *Paraceratherium* (which they called *Baluchitherium*). Four men were required to carry the femur alone of this enormous beast.

RIGHT: A life restoration of *Paraceratherium*, which reached an extraordinary twenty-six feet in length. *BELOW:* Following the discovery of the *Paraceratherium* skeleton in a badland gully, the entire team pitched in to unearth and remove the gigantic specimen.

1919: MEMO FROM THE GROUCHLESS GANG

Even as he began his first foray into the Gobi Desert, Roy Chapman Andrews was filled with the ebullience that would characterize his expeditions over the next eleven years. Here, from his unpublished journals, are the plans and specifications for the 1919 journey.

OBJECT: To get to Urga eventually   **MOTTO:** We should worry

PERSONNEL

Mr. R. C. Andrews—"Gobi"
  Head cook, skinner, butcher and general camp arranger and grouch
Mrs.—ditto—"Gobina"
  Photographer, Assistant cook, Meal and table arrangements
Mr. Mac Callie, alias "Delco"
  Chief Electrician, tent pegger, Water purveyor and wood cutter
Mrs. Mac Callie "Delcette"
  Coffee, tea and soup supply chef, table linen and cutlery
Mr. C. L. Coltman "Boss"
  Motor Engineer, time keeper, argot expert, and general commander
Mrs.—ditto—"Bossene"
  Assistant cook, quartermistress, finder of lost articles
Mr. Owen "Uncle John"
  Assistant Motor Engineer and all round help-less

REGULATIONS

1. No cussing the weather.
2. No insinuations if there is sand in the soup.
3. No grouching against the gasoline in the drinking water.
4. No profanity unless of picturesque variety.
5. All hands assist at unpacking and packing in evening and morning stops and starts.
6. All male members must take share in pumping tires and of work requiring more than hot air.
7. Camps will be made, starts made, stops made, and such disarrangements by vote, four votes carrying the day.
8. Any breach of regulations will be considered by court after dinner and during smoking hour (when most lenient treatment can be hoped for) and penalty judged will be walked by the culprit in miles recorded by speedometer at the start of the following day.
9. If male members of expedition cannot supply fresh meat on any one day they will not be allowed to smoke after dinner.

PLANS

1. To have a thoroughly good time
2. To get good specimens of all game available
3. Camp early and start late on general principle
4. To stop and investigate, or leave the road and explore whenever desired

(Signed) The grouchless Gang

LEFT AND OPPOSITE: Andrews and a gallery of Mongolian portraits: Curious about the explorers' quest for fossils, the nomadic Mongols were fascinated by cameras, phonographs, and other devices. On the other hand, Andrews and his team spent many a day cursing modern machinery's inability to withstand harsh desert conditions. Sometimes, the simplest way—sketching map routes in the sand, carrying spare tires by camel—proved to be the best.

The magnificently beautiful and eerie Flaming Cliffs of Shabarakh Usu.

1923: THE FLAMING CLIFFS AND BEYOND

The first exploration of the Gobi Desert may not have lived up to Roy Chapman Andrews' or Henry Fairfield Osborn's high expectations, but by all other estimations, it had been a great success. As a result, even more was expected of the second expedition, and the assemblage of collectors aboard the 1923 caravan reflected the increased hopes. Walter Granger was still chief paleontologist, but with him came such experienced prospectors as Peter Kaisen (Barnum Brown's assistant on many successful trips), George Olsen, and Albert Johnson.

The region had grown no safer, however, in the year since the last expedition, and this party was also forced to brave a string of dangers. Bandits and renegade soldiers were abundant along the caravan routes, and just days before the expedition headed into the desert, a pair of Russian cars had been waylaid and a man killed by Chinese soldiers in an area Andrews hoped to travel.

To offset these dangers, the expedition contracted for a military escort through several unsettled regions. Andrews realized the importance of an escort when he left the party for a few days in search of supplies and was attacked by bandits on horseback.

"[K]nowing that a Mongol pony never would stand against the charge of a motor car, I instantly decided to attack," Andrews recounted. "The expected happened! While the brigands were endeavoring to un-ship their rifles which were slung on their backs, their horses went into a wild series of leaps and bounds, bucking and rearing with fright, so that the men could hardly stay in their saddles." When he last saw them, Andrews said, the bandits "were breaking all speed records on the other side of the valley."

ABOVE: George Olsen, Albert Johnson, and Kan Chuen Pao (called "Buckshot"): three of the expert fossil hunters who accompanied the 1923 expedition.

Once reunited, the expedition plunged on. The first stop was Iren Dabasu, site of the initial discoveries of dinosaur bones the previous summer. Here, the collectors took a month to explore further, finding an enormous trove of Cretaceous fossils, including masses of hadrosaurs, or duck-billed dinosaurs.

Next, the expedition made a brief stop twenty-three miles south at a camel stop called Irden Mannah, where one of the party's Chinese assistants (Kan Chuen Pao, known as "Buckshot") found the skull of the remarkable Eocene mammal *Andrewsarchus*.

In spite of such exciting finds, Andrews was eager to return to the Flaming Cliffs of Shabarakh Usu and to leave the last vestiges of civilization behind. "All of us had the feeling that the expedition was just starting when we left Irden Mannah," he wrote in his journal. "From that time communications with the outside world ceased; we would be almost as isolated as tho' we were at the North Pole."

The route through the desert toward Shabarakh Usu left no doubt that the expedition had journeyed far from the modern world. "The surface of grey, brown and deep black was hardly relieved by a touch of green; not a sign of human habitation for many miles, not a flock of sheep or even camels," Andrews wrote. "A region of desolation, a land God forgot!"

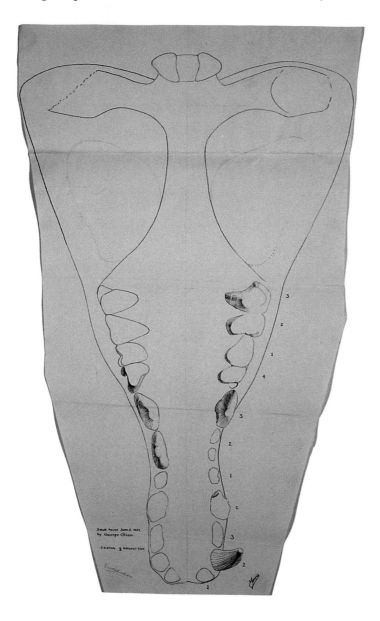

ABOVE: F. K. Morris' initial sketch of the jaw of *Andrewsarchus*, the bizarre Mongolian mammal found by George Olsen in 1923.

THIS PAGE: *Protoceratops*, the primitive ceratopsian, or horned dinosaur, didn't have an actual horn, merely a thickened bump on its snout (see skull at left). Andrews' team found *Protoceratops* skulls and skeletons along with nests containing eggs laid in concentric circles (above left), which were the first dinosaur eggs ever discovered.

The group reached the Flaming Cliffs on July 8, and for the next five weeks, they prospected this famous, wonderfully rich site. In less than a day, the collectors had found many remains of a previously unidentified ceratopsian dinosaur, including several skulls. By the end of day two, the haul was ten skulls as well as many other bones from the dinosaur subsequently named *Protoceratops*.

Perhaps the most exciting discovery of all took place on July 13, when assistant paleontologist George Olsen returned to camp reporting he had found some fossil eggs. Andrews and Granger were skeptical at first, thinking that Olsen had most likely found some stone concretions, which can mimic the size and shape of eggs.

Seeing the find themselves, however, the senior scientists were convinced: Olsen had unearthed the first eggs ever to be positively identified as belonging to a dinosaur. The scientists then realized in hindsight that the eggshell fragment they had found the previous year and assumed was from an ancient bird was also a dinosaur eggshell.

Due to the abundance of *Protoceratops* remains in the area, the scientists hypothesized that the eggs must have belonged to these ceratopsian dinosaurs. Later research has confirmed this estimation.

Almost as exciting as the eggs themselves was the undisturbed condition in which they were found. Many were unearthed in the same position they had been laid tens of millions of years earlier: in nests containing several eggs deposited in concentric circles.

Not that the *Protoceratops* skeletons and eggs were the only discoveries in the Flaming Cliffs area. The first described specimen, which is called the "type," of the small predatory dinosaur *Oviraptor* was found actually lying atop a clutch of *Protoceratops* eggs, leading Granger and Andrews to guess that the predator had been robbing the nest at the moment both the dinosaur and eggs were killed, most likely by a sandstorm. Other dinosaurs whose type specimens were found at the site include the fleet-footed predators *Saurornithoides* and *Velociraptor* and the armored dinosaur *Pinacosaurus*.

Without doubt, the 1923 expedition stands as one of the most fruitful of all time, and the discovery of the dinosaur eggs gained extensive newspaper and magazine coverage across the United States. Unfortunately, the discovery also sowed the seeds of discontent that eventually brought Andrews' expeditions to an end seven years later—and helped keep western scientists out of the Gobi Desert for more than sixty years.

TOP: George Olsen and Andrews examine a *Protoceratops* nest.   *ABOVE:* Tens of millions of years earlier, *Oviraptor* may also have examined just such a nest; the first *Oviraptor* skeleton ever found was lying atop a clutch of eggs.

Back in New York City that winter, Andrews used the publicity he had garnered to raise $280,000 for the next expedition. As part of his campaign, he allowed one of the dinosaur eggs to be auctioned, a publicity stunt that resulted in a five-thousand-dollar sale. The Chinese and Mongol officials saw this lucrative sale as evidence of each egg's worth and accused Andrews of removing valuable resources form the Gobi. From then on, the Museum expeditions were the objects of ever-increasing suspicion.

1925 TO 1930: THROUGH A SEA OF TROUBLES

Andrews and his team returned to Mongolia three more times after 1923. The 1925 expedition was elaborately planned and well financed, but suspicious local officials had begun to interfere with the work, throwing up logistical roadblocks that threatened to prevent the party from visiting the regions they wished to explore.

Still, the 1925 journey is remembered for a series of very important (if not famous) finds. These discoveries actually began when close examination of the bones brought back in 1923 revealed that some of those collected at the Flaming Cliffs belonged not to dinosaurs but to a tiny mammal dating from the Cretaceous.

Scientists had long known that small mammals coexisted with the dinosaurs, but fossil remains of such ancient mammals were nearly nonexistent. Returning to the Flaming Cliffs in 1925, Granger and his men soon found the skulls of seven more Cretaceous mammals.

These specimens included both placental mammals (those mammals, like modern dogs, cats, and humans, that give birth to fully developed young)

and marsupials (those mammals, including kangaroos and opossums, that give birth to less fully developed young that must spend time in their mother's pouch). Both types of mammals occurred in the same rock strata and had therefore shared the same landscape of the ancient Gobi with *Protoceratops, Velociraptor,* and the other remarkable Mongolian dinosaurs found by the Museum collectors.

The 1925 expedition was Andrews' last truly successful foray into Asia. Antiforeign sentiment, fanned by a nationalist group called the Society for the Preservation of Cultural Objects, was growing in China, and permits to travel and prospect in Mongolia were becoming more and more difficult to obtain.

In 1928, the scientists did make it back into the field, unearthing dinosaur eggs in Iren Dabasu (site of the first discovery of fossils during the 1922 expedition). But upon their return to Beijing, their finds were confiscated by Chinese authorities. The Museum retrieved its specimens only after intensive, bitter negotiations that lasted for weeks.

Clearly, conditions in the Gobi were becoming untenable. After the 1930 expedition (during which the explorers found a mass grave of the shovel-tusked mastodon *Platybelodon*), Henry Fairfield Osborn decided that the expeditions were no longer worth the risk or expense. Andrews closed down his headquarters in Beijing, and the great era of the Central Asiatic Expeditions came to an end.

Andrews and his team of explorers never found the human missing link Osborn had hoped for, but what they did bring home was so spectacular that few remembered that the search for human origins had been the expedition's initial goal. As W. D. Matthew expressed in a letter to Walter Granger, "The last word about your find is more than magnificent. Words fit to cope with its splendor have not yet been invented."

For anyone prospecting in the Gobi, packing and crating the specimens are just the beginning. Next the crates had to travel—by camel or truck —hundreds of miles to civilization.

The unmatched grandeur of the Gobi Desert.

THE 1930s AND BEYOND:
DISCOVERY ON A SMALLER SCALE

In regard to your request for further funds to investigate these localities, I am a little embarrassed. We just do not have funds available.
George Gaylord Simpson, in a 1947 letter to Barnum Brown

Roy Chapman Andrews' Central Asiatic Expeditions marked the last of the great paleontological forays sponsored by the Museum. The institution was hard hit by the Great Depression, as was every other major museum in the United States and abroad, shrinking endowments that had once funded expensive digs. In addition, when unrest began to grow in Europe, Asia, and elsewhere in the early 1930s, thought of any ambitious, far-flung expeditions was curtailed until

LEFT: The spectacular landscape of northwestern New Mexico, home to artists, miners, Native American trading posts—and dinosaurs.   *INSET:* The headquarters at Ghost Ranch, site of some of the Museum's most important discoveries of the 1940s.

after conditions improved—which, of course, didn't occur until well after the end of World War II.

But such outside forces as war and economic depression were not the only reasons for the Department of Vertebrate Paleontology's straitened collecting efforts; times were changing within the department as well. Perhaps the biggest change came in 1932 when Henry Fairfield Osborn, who had been devoting more and more time to his controversial (and frequently repellent) theories about race and the evolution of man, retired as president of the Museum and then died three years later.

At the height of his tenure, Osborn had promoted his belief in the necessity of ongoing new discoveries, new restorations, and new displays to please a fickle and easily bored public. With his retirement, partly in reaction to the virtual elimination of Museum funds for fieldwork, the Department of Vertebrate Paleontology entered a period of drift from which it did not completely emerge for many years.

The trustees must have realized that the Museum as a whole was losing its luster, for in 1934 they asked Roy Chapman Andrews to take over as director (at that time the most powerful position at the Museum). Andrews accepted the post, but it soon became obvious that he was not suited to handle the day-to-day responsibilities of running the Museum's labyrinthine bureaucracy.

With the Museum reeling from the twin blows of outside political and economic events and a rocky internal transition, it is doubtful that any Museum director could have moved forward with major achievements at this time. For Andrews, who had built his image as hunter, explorer, and adventurer so painstakingly, facing the mundane, yet often insoluble, problems associated with the operation of a huge institution was a nightmarish task.

The director's unhappiness and frustration are easy to detect in a portrait painted while he was in office. Looking stiff and unhappy in a gray suit, gray tie, and monocle, he seems a world away from the khaki-clad expedition leader whose journals brim with such high spirits.

Andrews himself must have realized that the portrait revealed more than he desired, because upon his departure from the Museum in 1941, he ordered it destroyed. His orders, however, were disobeyed, and today the painting hangs in the special collections department of the Museum library, a vivid reminder of a dark time in the history of the American Museum.

TRACKING THE GREAT SAUROPODS

During these rocky years, the Department of Vertebrate Paleontology did manage to mount a variety of expeditions into the field, sending collectors to bring back fossil mammals and reptiles from Florida, Arizona, and elsewhere. Perhaps the most important collecting efforts of the 1930s were those undertaken in Texas by collector and preparator Roland T. Bird (1899–1978).

Starting out as an independent fossil hunter who rode from site to site on a Harley-Davidson motorcycle equipped with a sidecar, Bird worked as Barnum Brown's assistant during several field seasons in the mid-1930s. Today, however, he is remembered for the discoveries he made on an expedition he undertook on his own to Glen Rose, Texas: the most spectacular and best-preserved dinosaur trackways (series of footprints) found up to that time.

Bird frequently acknowledged how much luck was involved in his discovery of the Glen Rose trackways. His story always began in 1938, when he happened upon an Indian trading post in Gallup, New Mexico, and found inside some blocks of stone containing what looked like the three-toed footprints of a dinosaur. Asking a trader where the prints came from, Bird was told that they had been shipped from Glen Rose, a small Texas town on the Paluxy River, southwest of Fort Worth.

This was the first Bird had heard of Glen Rose and its possible paleontological riches. What he and subsequent Museum scientists and historians never learned, however, is that the Museum had received information on the trackways in 1924—a full fourteen years before Bird made his lucky visit to the trading post—but had bungled the opportunity to collect specimens at that time.

In March 1924, the Department of Vertebrate Paleontology received a postcard from a man named E. G. Sicard, who was taking a cure in a Glen Rose sanitarium. On the front of the card was a photograph of what seemed to be the footprints of a large meat-eating dinosaur. The postcard bore a caption: "DINASAUR TRACKS, GLEN ROSE, TEX." On the reverse side, Sicard wrote, "This card may be of interest to you."

It certainly was. Dinosaur trackways were rare and almost never found in good condition. "If these are genuine tracks, they are remarkably fine ones," curator Walter Granger wrote to Sicard in April. "Curator Matthew is going to Texas on a reconnaissance trip in May and it is possible that he might be within reach of Glen Rose at some time during his wanderings."

W. D. Matthew did, in fact, visit Texas during the spring of 1924. He brought along a young assistant named George Gaylord Simpson, who was destined to become one of the best-known and most prolific Museum paleontologists of the post-Osborn years.

The 1924 trip, Simpson's first for the Museum, was an eventful experience for the young scientist. "One of Matthew's oddities… was that he had not learned to drive a car," Simpson recalled in his memoir, *Concession to the Improbable*. "A requirement, in fact his major requirement, for a field assistant was to operate and maintain a Ford Model T. In fact I had never driven a car either…."

But this lack of experience didn't stop young Simpson from claiming to be a fine driver, even after he got behind the wheel of the Model T. "Off we went somewhat jerkily into the traffic, and I believe that Matthew never realized how close he came to ending his career at that high point."

OPPOSITE PAGE: Found and lost: the misplaced postcard that alerted Museum paleontologists to the splendid fossil sites in Glen Rose, Texas, in 1924—fourteen years before Roland T. Bird stumbled on the same sites. *ABOVE:* Roland T. Bird with his greatest discovery: the trackways of sauropod and carnosaur dinosaurs.

returned often to South America, using its fauna as a springboard for his wide-ranging analyses of evolution and other subjects.

Simpson's early fieldwork also helped him hone the witty, frequently acerbic style that marked much of his writing. In a 1934 letter to Walter Granger, for example, he described the Patagonian custom of giving elaborate farewell parties. "They say here that no fiesta is good without at least one hospital case as a result," he wrote. "[A]fter that party one man went to the morgue, one to the hospital and one to prison, so it was a roaring success."

Unfortunately, one of Simpson's later visits to South America led indirectly to his departure from the Museum. During an exploration of the Brazilian Amazon in 1956, the paleontologist was struck by a falling tree, which shattered his right leg. It took eight days by canoe and airplane to bring him to a hospital in the United States, and two years and twelve operations passed before Simpson was finally able to walk again, with the help of a cane.

When he finally considered himself ready to return to work, Simpson learned that the Museum would no longer allow him to undertake any journeys into the field. Rather than accept this limitation, he resigned immediately from the department and continued his productive career (including worldwide travel) for another twenty years under the auspices of such institutions as Harvard and the University of Arizona.

Edwin H. "Ned" Colbert (born in 1905) made it clear in *Digging into the Past* that he did not place himself in George Gaylord Simpson's intellectual class. Whatever the truth of this modest self-assessment, no one can doubt that Colbert's contributions to the Museum's dinosaur halls rank among the greatest of any collector since Barnum Brown.

Colbert's most famous accomplishment by far was his team's 1947 discovery of a vast assemblage of skeletons of the Triassic dinosaur *Coelophysis* at Ghost Ranch, New Mexico. Perhaps the most important paleontological event in the 1940s, this find added immensely to our hitherto scanty knowledge of the primitive animals that inhabited the earliest phase of the age of the dinosaurs.

Colbert had ascertained that northern New Mexico was one of the best places on the continent for finding exposed rock dating from Triassic times. To the layperson, the jagged cliffs and towering rock formations in the area are all part of the breathtakingly beautiful landscape, but to the practiced eye, the rock strata running in clearly delineated stripes across the cliffs tell another story. Here, on a single mountainside, one can view exposures of the Triassic Chinle

Working the *Coelophysis* site at Ghost Ranch.

Margaret Colbert's restoration of the living *Coelophysis*: agile, alert, intelligent.

Formation; the Jurassic Entrada, Todilto, and Morrison Formations; and the Cretaceous Dakota Formation—sampling the entire reign of the dinosaurs at a glance.

Colbert knew that Triassic fossils had been found in the Chinle badlands near Ghost Ranch, and it took only four days of searching before his assistant, George Whitaker, found evidence of exactly how rich the area's Triassic beds were. Whitaker returned to camp carrying samples from a bone bed he'd found eroding out of a hillside, a bed that appeared to contain Triassic fossils in abundance. "They were an electrifying sight, those few bone fragments in George's hand," Colbert wrote. He recognized immediately that the fossils belong to *Coelophysis*, one of the earliest of all dinosaurs, which had previously been known only from fragmentary material.

During the rest of the summer and in 1948 as well, Colbert's team excavated dozens of complete *Coelophysis* skeletons from Ghost Ranch. The *New York Times*, *Life* magazine, and many other publications covered the find, and several of the best *Coelophysis* specimens are among the highlights of the Museum's dinosaur halls.

Today, Ghost Ranch serves as a conference center for the Presbyterian Church. On the grounds is the Ruth Hall Museum of Paleontology, which is open to the public and contains fossils of

TOP, LEFT AND RIGHT: Retrieving *Coelophysis* specimens from a quarry filled with dozens of skulls and skeletons of the early dinosaur. *BELOW:* The skull and long, toothy jaws of *Coelophysis*.

Coelophysis and less familiar Triassic creatures, memorabilia, and other treasures. The centerpiece of the small museum is an eight-ton block of stone, still being worked on by paleontologists, which harbors yet more specimens of the little dinosaur that must have lived in abundance here more than two hundred million years ago.

Ned Colbert also participated in a historic 1969 expedition to Antarctica in search of fossils. Because no ancient animals could have survived on the continent's frozen wastes, the paleontologist knew that if fossils turned up, they would provide evidence that Antarctica had once been located far to the north and had gradually been carried southward by the movements of the earth's tectonic plates, confirming the then-controversial theory now widely known as continental drift.

Colbert and the other researchers didn't have to search for long. On November 23, their first day exploring a series of cliffs called Coalsack Bluff in the Central Transantarctic Mountains, they came upon fossil bones dating from Triassic times—bones similar to those found in South African fossil beds of similar age.

The scientists immediately recognized the significance of their find. "We *did it!*" wrote a jubilant Colbert in a November 24 letter. "This really pins down continental drift, in my opinion," he added, signing his letter, "A very excited old man."

Within two weeks, the team had unearthed the bones of a variety of Triassic creatures that had also lived in Africa. Perhaps most important were the abundant remains of an animal named *Lystrosaurus* (a dicynodont, an early relative of mammals), one of the characteristic inhabitants of Triassic sediments in South Africa.

OPPOSITE: The barren, beautiful Antarctic landscape—which Edwin Colbert thought would be full of fossils. *OPPOSITE, INSET:* Colbert (at right), with Dr. David Elliot of Ohio State University, examining the vitally important fossil find at Coalsack Bluff in Antarctica. *ABOVE, LEFT AND RIGHT:* In letters home, Colbert acknowledged that Antarctica's strenuous conditions took their toll on him. But none of the challenges of living, traveling, or prospecting this remote area prevented his visit to Coalsack Bluff from being one of the highlights of his fossil-hunting career.

As Colbert pointed out, not only couldn't *Lystrosaurus* have survived the harsh conditions of Antarctica in its present location, but the dicynodont also could never have made the long oceanic crossing from Africa to the polar continent. Apparently, then, Antarctica had in fact been located far to the north, forming a semitropical supercontinent with Africa, and had been carried southward by the slow movements of the earth's tectonic plates.

Ned Colbert retired from the Museum in 1970, and even today he continues to work at the Museum of Northern Arizona in Flagstaff. Still adding to his impressive list of published works, in 1993 he produced his latest book, a biography of W. D. Matthew.

In many ways, Colbert was a transitional figure in the department's history, the last paleontologist to be brought on board by such early giants as Matthew, Walter Granger, and even the great Henry Fairfield Osborn himself. Colbert, however, also worked alongside a new generation, individuals such as Richard Tedford and Malcolm McKenna who today are bringing the department into a new era of active fieldwork and exploration.

THE 1990s:
BIRTH OF A NEW ERA

Of course we could spend all our time managing the collection. But I think a department loses its spark, its creativity, if it doesn't make fieldwork a very important part of its mission.
—Richard Tedford, chairman of the
Department of Vertebrate Paleontology, 1993

A t times, the extraordinarily vivid and productive history of the Department of Vertebrate Paleontology seems to hang heavily over the present-day staff. The era of Henry Fairfield Osborn, Barnum Brown, and Roy Chapman Andrews is long past, and today's Museum paleontologists are well aware of how undramatic much of their research seems when compared to the time when steamships and railroad freight cars brought tens of thousands of pounds of fossils to the Museum from places with such exotic names as Hell Creek, Bone Cabin Quarry, and the Flaming Cliffs of Shabarakh Usu.

Mongolia's vast and forbidding Gobi Desert—as well as countless other fossil sites in North America and throughout the world—remain fertile destinations for today's paleontologists.

Until the renovation of the vertebrate paleontology halls began in 1991, the spirits of the past seemed to inhabit every dinosaur, early mammal, and other fossil animal on display. It was impossible to look at Brown's "favorite child," his *Tyrannosaurus rex*, without thinking about his unparalleled role in building the strongest collection of vertebrate fossils on earth or to view the magnificent *Protoceratops, Paraceratherium,* or *Andrewsarchus* without recalling Andrews' matchless adventures in Mongolia.

The new halls, of course, pay homage to these men and the other creators of the Department of Vertebrate Paleontology, but the vivid memories have inevitably begun to fade. Perhaps it is because their ghosts don't coexist easily with the shiny new dinosaur models, the interactive computer programs, and the screens running endlessly looped videotapes. Although the past's shadows have retreated, they haven't vanished entirely. Today they survive in the Frick Building, the ten-story home of the Department of Vertebrate Paleontology that sits inside the Museum's inner courtyard, hidden from public view. Built partly from funds given by wealthy paleontologist Childs Frick, who also donated much of the Museum's magnificent fossil mammal collection, this building holds the departmental offices, the mounting and preparation laboratories, and perhaps most importantly, the treasured results of a century and a quarter of active fieldwork.

Thousands of specimens, undoubtedly including many vitally important finds, await study in the Museum's Frick laboratory and storage areas.

Nearly every floor of the Frick Building is crowded with tall cabinets of greenish metal. Inside these "cans" (as some of the scientists call them), and frequently on top of and next to them, are bones of ancient mammals: mastodon jaws, camel femurs, carnivore skulls, horse toes and teeth. Tens of thousands of individual fossils, sorted and labeled and stockpiled for future research or display.

Not all the fossils have been so carefully classified. One large wooden chest on the fifth floor contains an entire mummified ancient bison, brought back from Alaska. Other rough-hewn wooden crates hold what look like white boulders. These lumpy objects contain specimens that are still encased in the protective plaster—called "field jackets"—they were packed in at the excavation site. Such specimens, dating from as long ago as the 1940s, their plaster often scrawled with the handwriting of the men who found them, provide a particularly poignant link to the department's past.

While the mammal storage areas are brightly, almost harshly lit by banks of fluorescent lights, the basement dinosaur storerooms cast an entirely different spell. Dingy, windowless, these rooms are cluttered with dusty shelves that reach almost to the cobwebbed ceilings. On every shelf lie precious pieces of the Museum's dinosaur collection—those pieces of fossilized remains that haven't yet warranted preparation for display in the dinosaur halls.

One small room is dominated by *Allosaurus* remains, including vertebrae and foot bones labeled "Como Bluff" and "Bone Cabin Quarry 1901." This room also boasts a battered cardboard box filled with fine orange sand, a memento of Andrews' Central Asiatic Expeditions. Another, even smaller annex contains more of Mongolia's treasures: *Protoceratops* eggs as well as some unidentified dinosaur eggs that were also found by Andrews' team.

The largest storeroom is divided into Saurischian and Ornithischian sections. Here are enormous sauropod bones, including thigh bones that seem incomprehensibly huge in such modest surroundings. The skull of a *Protoceratops* can be found here too, with bones protruding like mysterious whitish growths from orange Mongolian stone. Also in this room is a radioactive dinosaur bone, unearthed from a radium deposit, which at one time was accompanied by a note: "Perhaps Madame Curie would be interested."

As Richard Tedford points out, it would be easy for today's staff members to spend the rest of their careers sorting through the countless thousands of specimens brought home by the fossil hunters of an earlier age. And in fact, some paleontologists do devote a portion of their time to previously discovered specimens.

TOP AND ABOVE: Dinosaurs and other ancient creatures seem to come alive in the Museum's dark and ghostly storerooms.

A specialist in ancient mammals, Bryn Mader, for example, is well aware that Henry Fairfield Osborn and other early scientists sometimes allowed their enthusiasm to overwhelm their common sense. As part of his responsibilities, Mader has sought to determine exactly how many different types of oreodonts (a diverse group of mammals related to both modern pigs and deer) the Museum actually harbors.

Bending over a row of skulls placed atop one of the storeroom cans, Mader points at two of them. "These were both identified as type specimens of different genera, but they look pretty similar to me," he says. "They come from the same locale, the same strata— the only real difference seems to be that this one was crushed, so it has a slightly different shape. I think they actually belong to the same genus."

Mader believes that careful, systematic review would help reveal many incorrect classifications made by earlier paleontologists. Such a review would also lead to the identification of some new "discoveries," previously unclassified genera that were either never scientifically described or that were originally incorrectly identified.

Senior scientific assistant John Alexander is engaged in detective work of a different kind. As manager of the Museum's fossil mammal collection, he has undertaken a years-long project to untangle the collection's often snarled records. In the process, he discovered that many vitally important specimens that had been lent out to other institutions or to individual scientists had never been returned. Before Alexander retrieved them, some had been in outside hands for generations.

As Tedford also makes clear, however, few of today's paleontologists would be content to work entirely within the confines of the Frick Building. In fact, when describing how they became interested in paleontology, most tell remarkably similar stories, and all the stories begin with childhood fossil-hunting expeditions. "We grew up reading about the great dinosaur hunters," Tedford says. "We wanted to make discoveries as important as theirs, and that required fieldwork."

Fortunately, after many years of sporadic, often scant financial support, the Department of Vertebrate Paleontology today is comparatively well funded. "We're fortunate that we have a variety of special endowments to draw on," says Tedford. "So even if outside support is hard to find, we can still administer a scholarship program, support our graduate students to some degree, and also undertake a certain amount of fieldwork."

ABOVE: Paleontologist John Alexander surveying the barren Wyoming landscape that, during the Eocene period, hosted dense forests—and the early primate *Notharctus*.   *LEFT: Notharctus* was an agile and energetic animal whose long, grasping fingers and toes made it ideally adapted to life in the trees.

Most of the Museum's current field projects, which are centered in Wyoming and Montana, are of a small scale: "Dirty deeds done dirt cheap," as John Alexander calls them. But even these modest efforts are helping fill in the gaps in our knowledge of the ancient fauna of North America—and on occasion, they produce a vitally important find.

A HORIZON FULL OF *NOTHARCTUS*

One such find took place in June 1988 in the badlands of southwestern Wyoming's Bridger Basin, where department curator Eugene Gaffney and senior scientific assistant Frank Ippolito were hunting for Eocene fossil turtles. "I was scrambling up a hill on all fours, trying to take a shortcut," Ippolito recalls, "when, just a few inches from my face, I spotted some fossil fragments sticking out of the rock."

Ippolito and Gaffney excavated a block of stone that they hoped would contain more of the fossil. In removing some of the stone to lighten the block, they exposed portions of the skull and skeleton of a small mammal they did not recognize.

Back at the Museum, the largely unprepared specimen was first identified as *Hyopsodus,* a small relative of modern hoofed mammals whose remains are frequently found in the Bridger Basin deposits. If the specimen had been put in the storeroom among the Museum's other *Hyopsodus* fossils, it probably would have stayed there for years. Instead of storing the skeleton, however, John Alexander placed it on his desk. He wanted it at hand so he could try to figure out why it interested him so much. After several weeks, he took it to a nearby lab and used a tiny pneumatic hammer to separate the specimen's bones from the surrounding rock that obscured it.

Alexander knew immediately that this was no *Hyopsodus.* "Its skull had a postorbital bone around the eye," he says. "This bone occurs in very few mammals—but it does occur in primates. It turns out they had found one of the best-preserved skulls of a primate ever discovered in North America."

Alexander soon identified the specimen as belonging to a male *Notharctus,* a primate long considered to resemble modern lemurs (the unique primates found only on Madagascar). However, the completeness of the specimen—as well as another specimen, which

had been misidentified as a carnivore, that Alexander found in the Museum's Eocene mammal collection—showed that previous restorations of this important primate were incorrect in many physical details. Perhaps most importantly, *Notharctus* shared several previously unseen features with ancient ancestors of modern monkeys, including grasping hands and feet.

Although the Bridger Basin today is composed of stark, dry badlands, during the Eocene period the region was covered in dense forest. There is no doubt that *Notharctus* was a tree-dweller, with strong hind limbs, long hands and fingers, and a long tail for balance, all of which enabled it to leap with agile grace from branch to branch. The physical adaptations of *Notharctus,* as well as many other creatures that shared its habitat, made them well suited to their forested habitat.

Returning to the Bridger Basin site in 1991, Alexander had trouble finding further evidence of the ancient primate at first, but then he came upon a second male *Notharctus* skull. Subsequently, a chance sighting of a tiny finger bone protruding from a rock face led to the quick discovery of yet another skull and partial skeleton, this time of a female individual.

Today, casts of Alexander's skeletal mounts of a male and female *Notharctus* can be found in the Museum's fossil mammal halls, while another is in the Hall of Human Biology and Evolution. Having prepared the little primate for display in both these new halls at the Museum, Alexander is eager to return and explore the Bridger Basin region further.

"This area, which is just off Interstate 80, has to be considered one of the best places in North America to find ancient primates," he says, a wistful tone entering his voice. "The exposures stretch for hundreds of miles. Just imagine what we could discover if we had the time and money to do a full-scale search."

THE 1990s: BACK TO MONGOLIA

Like so many other budding paleontologists, Malcolm McKenna, Frick curator in the Museum's Department of Vertebrate Paleontology, dreamed as a teenager of retracing Roy Chapman Andrews' footsteps across the great wilds of the Gobi Desert in Mongolia. "I remember reading Andrews' books, about his discovery of *Baluchitherium* and other animals, and deciding that I had to get there," he recalls.

THIS PAGE AND OPPOSITE: Despite new navigating technology and other advances, the 1990s American-Mongolian expeditions to the Gobi Desert in most ways closely resembled Roy Chapman Andrews' journeys of sixty years earlier.

Nor did McKenna keep such dreams to himself. As far back as high school, Priscilla Coffey, his future wife, shared his goals. "In her high school annual, a friend wrote that she was going to marry me and we'd go off to Mongolia," McKenna says. "I don't think either of us could have guessed it would take more than forty years."

The McKennas did manage a short visit to Mongolia at the height of the Cold War in 1964. "No officials wanted to talk with us, and no scientists were allowed to," McKenna says. "We got into the desert, where the fossils were, and learned a lot about potential logistical problems, but of course we weren't allowed to collect."

The barriers to a new scientific expedition to the Gobi were formidable. Many were the results of the endless political tensions between China and Russia, which flank Mongolia, tensions that were apparent during Andrews' explorations and only grew worse in the years that followed. In addition, Americans visiting the region were suspected of being representatives of the hostile United States government and were viewed, therefore, with suspicion.

There were also other, more localized problems that conspired to prevent the McKennas or any other Museum paleontologists from visiting the Gobi. "There's still a certain amount of resentment left over from the old days, resulting from Andrews' arrogant treatment of people who probably knew more than he thought they did," McKenna points out. "People have long memories."

Finally, in 1990, the great upheaval that brought down so many communist governments in Europe claimed another victim: the hard-line Mongolian government, the last roadblock to renewed exploration of the Gobi. Soon thereafter, a Mongolian diplomat visited the Museum and inquired whether the scientists might like to begin a new series of expeditions. Needless to say, he received an enthusiastic response.

The initial program, planned in conjunction with a team of scientists from the Mongolian Academy of Sciences headed by paleontologist Perle Altangerel, consisted of three expeditions over three years, with the first to take place in the summer of 1991. The Museum contingent was headed by paleontologist Michael Novacek, Museum vice president and dean of science, and also included paleontologists Malcolm McKenna (responsible for scientific planning), Mark Norell (the group's field leader, charged with determining the best sites to explore), Lowell Dingus (assigned to record the layers of sediment where fossils were found), and James Clark, an experienced field collector and postdoctoral fellow at the Museum.

A FOSSIL'S JOURNEY

For Mark Norell, Perle Altangerel, and the other paleontologists who discovered *Mononykus* and many other fossils in Mongolia, locating the specimens presented only the first of many challenges. Before the scientists could study these important fossils, they had to remove them from the earth and bring them home in good condition—a challenging job, and one with no guarantee of success.

With minor variations, the excavation techniques used by today's fossil hunters differ little from those employed by Barnum Brown and other early collectors. Having located a dinosaur fossil, usually by spotting a portion of exposed bone that has eroded out of the ground, the paleontologist must then try to determine the extent of the bone (or bones) that remains buried. To do this, he or she may use small picks and shovels to make a series of excavations charting the placement of the fossil in the earth.

The next—and most time-consuming—task involves freeing the specimen from the surrounding rock (called the matrix). If the fossil is located in crumbly rock or sand, only small picks and fine brushes may be necessary to free it. But many fossils are trapped in extremely hard matrix, requiring jackhammers, bulldozers, and other heavy machinery to remove. Obviously, preserving a fragile fossil while attacking the matrix with a jackhammer requires great care, and many prize specimens have been ruined during excavation by sloppy fossil hunters.

Once the surface of the bone has been exposed, the paleontologist may paint the fragile surface with hardeners such as glues or shellacs. The exposed section of bone will then be tightly wrapped in moistened toilet paper and packed in plaster-soaked burlap or plaster bandages, which will protect the specimen during further excavation, removal, and shipment.

Only once it has been wrapped in plaster will the fossil be lifted from the bed in which it has lain for tens (or hundreds) of millions of years. After it is turned over, the newly exposed side will also be wrapped in toilet paper and plaster, the final step before the now-unwieldy specimen is ready for shipment back to the Museum.

At the Museum, the plaster and other coatings are removed, and staff preparators may remove whatever matrix still adheres to the fossil and harden the cleaned specimen with further coats of glue. Now the paleontologist can finally get down to the job of studying the specimen: identifying it through comparison with similar specimens, describing it if it is a new genus or species, and attempting to reconstruct a skeleton if enough bones have been found.

The painstaking process of unearthing, preparing, and packing fragile, irreplaceable specimens.

In addition, Priscilla McKenna, who had accompanied Malcolm on many of his expeditions, was responsible for much of the logistical planning and for mapping the group's route using Global Positioning System (GPS). This high-technology mapping system, light-years ahead of anything Roy Chapman Andrews could have imagined, picked up signals from United States satellites to determine the exact positions of the explorers and important fossil sites.

RIGHT: Discovered during the 1990s Gobi expeditions: a *Protoceratops* skull still in the ground.

Planning the first expedition, the Museum team discovered that much else had changed since Andrews led his intrepid explorers into the Gobi. The expeditions of the nineties, for example, were carried out on a far more modest scale than the great Central Asiatic Expeditions of the twenties and thirties. Unlike Andrews, today's explorers were unable to dispatch vast camel caravans to carry thousands of gallons of gasoline and supplies to distant caches. Instead, the Americans and Mongolians traveled in Mitsubishi jeeps and Russian trucks, carrying as much as they could along with them, and replenishing water and other supplies at towns and oases along the way.

Even with all the changes, however, much had remained the same. "Polish, Russian, and Mongol scientists have been working in the Gobi for decades, but there are still endless sites to be explored," says McKenna. "We also found that many of the risks involved haven't changed."

During one expedition, for example, the group was bedeviled by a truck that was constantly breaking down. "Also, the driver of that truck kept doing wild things like taking off in it to explore on his own," McKenna recalls. "The rest of us would have to keep him in sight, or else our gasoline supply might disappear for good."

ABOVE: A fine *Velociraptor* skull found by the expedition. *RIGHT:* Unearthed by the 1990s Gobi team: the bones of *Estesia*, an unusual lizard which might have had a poisonous bite.

The rewards made the risks worth it, of course, just as they had for Andrews. Each season, the team explored a variety of sites, revisiting some of Andrews' old haunts. "We returned to the Flaming Cliffs each of the first two years," says James Clark. "It's a beautiful site—plus, there are still plenty of interesting fossils there, besides the usual *Protoceratops* eggs and skeletons."

Other destinations, with such mysterious names as Tugrugeen Shireh, Khermeen Tsav ("A wonderful area, except for the ticks and swarming flies," comments Lowell Dingus), and Khulsan, had never been explored by western scientists. Overall, the three field seasons in the Gobi have produced many fine specimens, including several of the powerful theropod *Velociraptor* (which starred in the book and movie *Jurassic Park*); the little-known *Erlikosaurus andrewsi*, a member of a group of dinosaurs called the segnosaurs ("slow reptiles"); a possibly venomous Cretaceous lizard called *Estesia mongoliensis*; and a large number of important mammals, including *Zalambdalestes*, a shrewlike animal that lived alongside the last dinosaurs. "We've brought home several hundred specimens, so clearly we're doing quite well," Dingus says. "Of course, after we finish studying and making casts of them, we'll return the original specimens to Mongolia."

By far the most famous and controversial find thus far is *Mononykus* ("One claw"), an extraordinary flightless bird first described in 1993 by Perle Altangerel and the Museum's Mark Norell, Luis Chiappe, and James Clark. With only four specimens collected so far (including one brought home by Andrews in 1923, which had languished in the collection under the name "bird-like dinosaur"), *Mononykus* is so unlike any previously known animal that it seems certain to spur debate for years to come.

About the size of a turkey, *Mononykus* shared many characteristics with other small theropod dinosaurs, including a long, bony tail, toothy jaws, and many other skeletal features. "When we first found it, we thought we'd found an unusual theropod," Norell says. "It wasn't until we brought it home and looked at it more closely that we realized we'd dug up a bird."

The researchers' original misperception is less glaring than it might appear. Most paleontologists now agree that modern birds are the only living descendants of dinosaurs, that they are in fact living dinosaurs. The most birdlike ancient dinosaurs are large-eyed, large-brained, fleet-footed theropods such as *Dromaeosaurus* and *Velociraptor*, animals that shared many characteristics with *Mononykus*.

Other characteristics of *Mononykus*, however, are unique among any fossil animal yet discovered. *Mononykus* was named for its strangest feature: its shortened forelimbs, each only three inches long and ending in a single, strong claw. These limbs, the scientists believe, were perhaps adapted for digging, most likely as an aid to finding food.

What finally convinced the scientists Perle, Norell, and their coauthors that

THIS PAGE: Expedition co-leader Perle Altangerel (center) and several of the specimens discovered during the 1990s explorations (top and bottom).

Mononykus was a bird are several crucial features, including a keeled breastbone, fused wrist bones, and a birdlike braincase. "It seems clear that *Mononykus* was more closely related to modern birds than was *Archaeopteryx*, the first ancient bird discovered," Norell states.

Researchers outside the Museum remain more cautious, however. They discount the skeletal similarities and wonder why *Mononykus*, which lived about seventy million years ago, could not fly when the supposedly more primitive *Archaeopteryx*, which lived seventy-five million years earlier, could.

The Museum paleontologists feel that such doubts miss the point. "We infer fossil relationships based on the animals' anatomy, not on when they lived," Clark points out. Adds Norell, "Perhaps the line that led to *Mononykus* lost the ability to fly, or perhaps flight evolved independently in the line that led to *Archaeopteryx* but not in the one that included *Mononykus*. We don't know which of these hypotheses is true, but that doesn't prevent us from recognizing a bird when we see it."

For Malcolm McKenna and the rest of the department scientists, such ongoing debates provide the best reason for continuing to return to Mongolia, even now that the initial three-year program has been completed. "Henry Fairfield Osborn thought he'd find answers when he sent Andrews into the Gobi, but instead he found more questions, just as we are today," McKenna says. He pauses, then smiles and speaks what might serve as the motto for all fossil hunters dating from the birth of the field of paleontology through the present: "Sometimes the questions are more important—and a lot more interesting—than the answers."

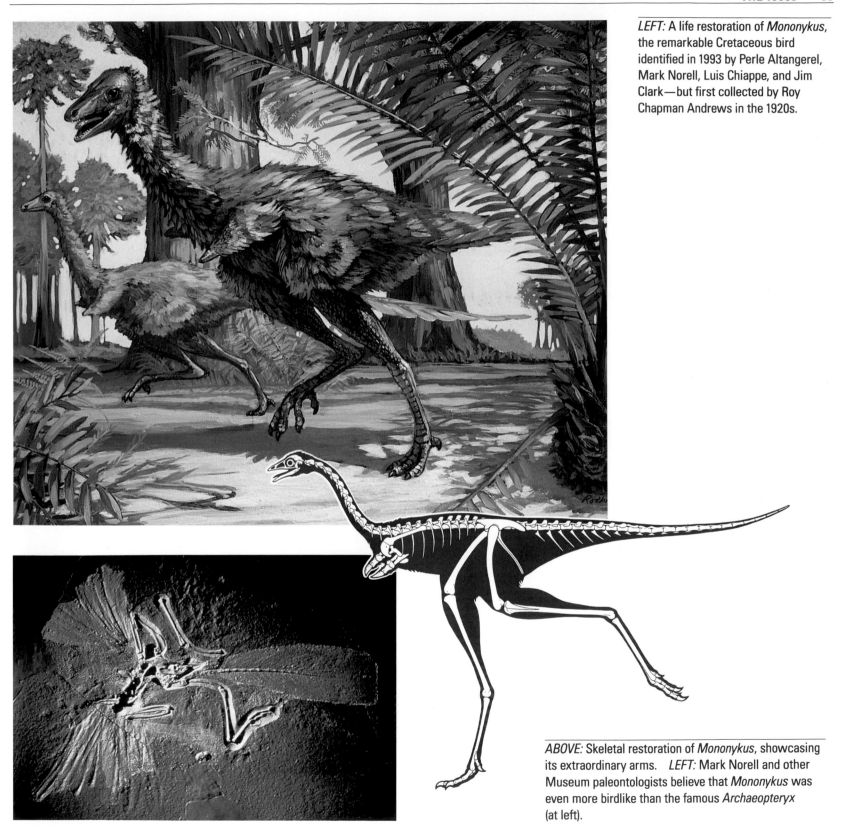

LEFT: A life restoration of *Mononykus*, the remarkable Cretaceous bird identified in 1993 by Perle Altangerel, Mark Norell, Luis Chiappe, and Jim Clark—but first collected by Roy Chapman Andrews in the 1920s.

ABOVE: Skeletal restoration of *Mononykus*, showcasing its extraordinary arms.   *LEFT:* Mark Norell and other Museum paleontologists believe that *Mononykus* was even more birdlike than the famous *Archaeopteryx* (at left).

PART TWO

THE
COLLECTION

HIGHLIGHTS OF THE NEW HALLS

Each year, more than 400,000 schoolchildren visit the Museum. If we can capture their imagination, we can give them a head start toward scientific literacy they otherwise might not have.
—Lowell Dingus, project director of the fossil hall renovation, 1993

By 1996 when the American Museum unveils the Hall of Primitive Vertebrates—the last stage in the Museum's forty-three-million-dollar plan to renovate its vertebrate paleontology halls and library—the public will have the chance to view the results of one of the most ambitious renovation programs ever undertaken by any museum.

Some of the changes are architectural. Where the old halls were dark and dusty, their windows painted over and their twenty-foot, scalloped ceilings hidden by faded

A mother and baby *Barosaurus* being menaced by a marauding *Allosaurus*: The first exhibit in the Museum's ambitious renovation of its dinosaur and other vertebrate fossil halls opened in the Roosevelt Rotunda in 1991.

plaster, the new design allows natural light to illuminate the galleries while also providing views of the surrounding neighborhood and Central Park.

The new halls also welcome modern technology to the traditional art of exhibit display. Alongside the fossil animals (which, cleaned and restored by the Museum's preparators, seem to glow with vitality) are informational video screens, interactive computer programs, and other devices that allow the visitor to learn more about many of the halls' most interesting denizens.

Most importantly, the new halls give the Museum's millions of visitors a chance to learn the most current information on the biology, habits, and evolutionary relationships of hundreds of ancient animals—or merely to see *Apatosaurus*, *Tyrannosaurus rex*, or other old favorites in a new light.

Those expecting a typical display of ancient animals—centered on a step-by-step journey through the history of life on earth, from the earliest to the most recent—are in for a surprise. "The first decision we had to make was whether to use a traditional 'walk through time,' or to base the design on evolutionary relationships," says Lowell Dingus, director of the renovation project. "We decided that, since the research here at the Department of Vertebrate Paleontology places a heavy emphasis on systematic relationships, we would go in that direction."

The emphasis Dingus is talking about is called cladistics, and for many biologists and paleontologists it represents the clearest method of studying the evolution of life over the course of hundreds of millions of years. Simply put, while traditional evolutionary theories put tremendous stock in when an animal lived, cladistic analysis is far more concerned about an animal's physical characteristics.

For example, as discussed in the previous chapter, traditionalists doubt that *Mononykus*, the fascinating Cretaceous animal that lived in the deserts of Mongolia, is a bird. They point to the fact that the flightless *Mononykus* lived seventy-five million years after *Archaeopteryx*, which could fly. In traditional evolutionary systems, this seeming devolution (from the ability to fly to a flightless state) rules out *Mononykus'* being a bird.

However, Mark Norell and others studying *Mononykus* have identified five discrete skeletal features, including a keeled breastbone and fused wristbones, that *Mononykus* shared with modern birds but *Archaeopteryx* lacked. To this group of scientists, these features show unequivocally that *Mononykus* was more closely related to living birds. When it lived and whether it could fly were

The delicate and complex mounting of *Moropus*, a huge relative of modern horses that inhabited the plains of central North America during the Miocene period, demonstrates the craft and art of fossil preparation from the age of Henry Fairfield Osborn to today.

far less important than how many physical characteristics it shared with its modern relatives.

According to senior project coordinator Melissa Posen, grouping the animals in this manner provides a challenging new way of looking at the history of life on earth. "Our computers and video terminals help explain why specimens have been grouped as they have," she says. "Also, the most important specimens are accompanied by three captions. One is targeted at the youngest and most casual visitors, the second to those with more than a passing interest, and the third to those looking for the most detailed information."

BUILDING THE PERFECT *T. REX*

Even after the painstaking jobs of locating and excavating a fossil specimen, shipping it home, and preparing and mounting it for display have been completed, the responsibilities of the Museum's experts may be far from over. Changes in scientific wisdom may render one of the mounted specimens obsolete years later, leaving the paleontologists no choice but to redesign the mount or to abandon it altogether.

Perhaps the most spectacular example of this process involves *Tyrannosaurus rex.* When it was mounted between 1914 and 1915, *T. rex* was placed in an upright posture, head set atop a stiff spine, tail dragging behind. This pose, familiar to anyone who has ever seen a Godzilla movie, represented the best estimation of the dinosaur's posture at the time.

In recent years, however, careful study of dinosaur bone structure has led scientists to conclude that the old posture was incorrect or at least incomplete. *T. rex,* they now believe, often walked or ran in a stalking pose: its body tilted far forward, its backbone nearly parallel to the ground, and its tail held stiffly for balance. This posture would have allowed the huge carnosaur to move with far greater speed, agility, and economy.

For today's preparators, headed by supervising exhibition assistants Jeanne Kelly and Phil Fraley, the challenge of dismantling and rebuilding *T. rex* proved enormous. "We can spend weeks, even months, cleaning and preparing a tiny mammal skull," points out Kelly, a preparator at the Museum for nearly twenty years. "We didn't have years to spend on *T. rex,* so we knew we had to work quickly—but also carefully."

Much of the initial responsibility fell on Fraley's shoulders. "We were dealing with a priceless artifact, something that was literally irreplaceable, so we knew we had to anticipate any possible problem as thoroughly as possible," he says. "Merely to get to the point of taking the skeleton apart safely for preparation and remounting took three years of planning."

The dismantling of *T. rex* was not without risk for the Museum staff, either. The dinosaur's pelvis alone weighed more than two thousand pounds and was suspended fourteen feet in the air on the original

RIGHT: Phil Fraley working on one of the hundreds of specimens being prepared for the Museum's new halls.

mount. One false move or miscalculation and the enormous bone could have fallen, with catastrophic results.

Once the skeleton was safely removed from the metal framework that had held it for more than seventy-five years, it was taken to Kelly's laboratory for cleaning and strengthening. "We use Zip-Strip to remove the old combination of orange shellac and alcohol from the bones, and then treat them with liquid plastic thinned with acetone," she explains. "This combination actually penetrates and strengthens the bone."

Fraley's next task was ensuring the structural stability of the metal framework (called an armature) designed to hold *T. rex* in its new stalking pose. "We needed to think about what would happen to the armature after a hundred years," he says. "Will the weight of the specimen create fatigue in the metal? What points in the armature will be under the greatest stress? We can't just guess at the answers— we have to know."

For Kelly, Fraley, and supervising exhibition assistant Steve Warsavage, stress of a different sort seems to be an integral part of their workday. "As soon as we're done with *T. Rex,* it's on to the next specimen, the next set of challenges," says Kelly. "But working on this collection is the chance of a lifetime, and I wouldn't trade my worries for anyone else's."

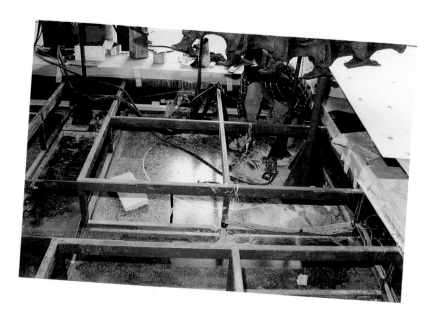

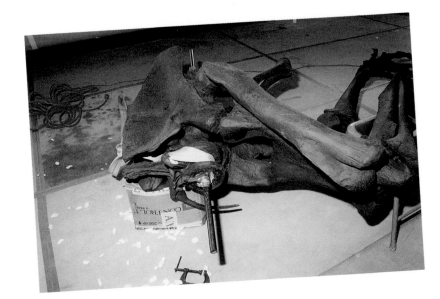

On the following pages you will find profiles of forty-three of the halls' most important denizens—ranging from such familiar inhabitants as *Tyrannosaurus* to an array of fascinating, little-known creatures. Along with detailing the animal's classification and when it lived, each profile provides a wealth of information on where the animal was found, how it made its way to the Museum, and what current scientific thinking leads us to conclude about how it may have lived.

For those who visit the new halls, these profiles should serve as a valuable souvenir. For those still waiting to enter these beautifully refurbished rooms, here is a small hint of what awaits you.

THIS PAGE AND OPPOSITE: As the last vestiges of the old halls were swept away, the ambitious new exhibits began to take shape. The exhibition areas were stripped and redesigned; fossil skeletons such as *Brontops* and *Megaloceras*, untouched for decades, were dismantled, cleaned, strengthened, and reconstructed; and casts of *Allosaurus*, *Barosaurus*, and others were built and installed.

GALLERY OF DINOSAURS

ALBERTOSAURUS
Order: Saurischia
Suborder: Theropoda
Infraorder: Carnosauria
Late Cretaceous
Discovered in 1914 by Barnum Brown and Peter Kaisen in the Red Deer River region, Alberta, Canada

One of the most productive field seasons of Barnum Brown's life was spent in the famous fossil-laden badlands of Alberta, Canada, during the summer of 1914. In a single letter to Museum curator W. D. Matthew, on June 24 of that year, Brown reported the discovery of specimens of *Albertosaurus* (then called *Deinodon*), *Ornithomimus*, *Corythosaurus*, *Monoclonius*, and *Ankylosaurus*—an extraordinary dinosaur haul.

Brown's 1914 season, however, was not unalloyed pleasure. Just a few miles down the Red Deer River, camped on the opposite bank, was another party of fossil hunters, sent by the Canadian Geological Survey and led by Charles H. Sternberg and his three sons. At the time, the elder Sternberg was Brown's chief rival for the fame associated with dinosaur hunting and had supplied many museums in the United States and abroad with his finds.

Although Brown couldn't deny Canada's inherent right to sponsor a prospecting team on its own soil, he still felt proprietary about the area, having been the first to explore the region, five years earlier. Early in the summer of 1914, tensions between the two camps rose, and Brown soon began to complain about the Sternbergs' behavior. "They have no regard for the ethics of bone digging," he wrote to Matthew, complaining that his rivals were excavating in areas he considered his own. Fortunately, the tensions at the 1914 site did not erupt into a full-scale confrontation. In fact, if anything, the rivalry inspired both groups to search even more aggressively and

ambitiously—and culminated in remarkable finds by both teams, which included the *Albertosaurus* whose skeletal plaque graces the Museum's new halls.

THE LIVING ANIMAL

Albertosaurus shared the Late Cretaceous plains of Montana and Alberta with many other familiar dinosaurs, including *Triceratops*. It may have hunted the hadrosaurs, such as *Edmontosaurus* and *Corythosaurus*, that were abundant at the time.

This twenty-six-foot-long carnosaur wasn't the only large predator to inhabit western North America; it seems to have lived at the same time as the far larger *Tyrannosaurus*. With that fact in mind, it is possible to imagine *Tyrannosaurus* chasing *Albertosaurus* off a kill, much as the more powerful lions do to cheetahs on the African plains today.

Albertosaurus, and its huge, powerful foot.

Anatotitan

ANATOTITAN

Order: Ornithischia

Suborder: Ornithopoda

Family: Hadrosauridae

Late Cretaceous

Two specimens on display: first specimen discovered in 1882 by collectors for Edward Drinker Cope near the Moreau River in South Dakota; second collected by Barnum Brown in 1906 in Crooked Creek, Montana

Barnum Brown, the Museum's premier fossil hunter, enjoyed telling anecdotes about his adventures, and the story of his "discovery" of the 1906 *Anatotitan* was clearly one of his favorites. More than seventy-five years later, the tale still brims with Brown's lifelong love of new finds and unexplored frontiers.

In both a history of Museum specimens and a 1908 issue of *Scientific American*, Brown

told of how two ranchers on horseback spotted the skeleton's backbone weathering out of a rock, many of the bones still in their original positions. Not recognizing the strange formation, the ranchers stopped and began arguing as to whether it was composed of actual rock or buffalo bones.

To settle the argument, one rancher dismounted and "kicked off all the tops of the vertebrae and rib heads above ground thereby proving by their brittle nature that they were stone and not buffalo bones," Brown, recounted. "The proof was certainly conclusive but very painful to the subsequent collectors."

Upon hearing the ranchers' story, a neighbor named Alfred Sensiba guessed that the formation might, in fact, be made of fossil bones and decided to acquire rights to the bones. Rather than paying cash, Sensiba came up with an acceptable alternative—and thus the Museum's future *Anatotitan* became what certainly must be the only dinosaur skeleton ever traded for a six-shooter.

At this stage of the story, Brown, who was fossil hunting in the vicinity, heard about the specimen and purchased the right to excavate it. As for the six-shooter, Brown claimed that it was in turn swapped for a pinto pony, which was appropriately named "Dinosaur."

THE LIVING ANIMAL

Until recently, the Museum's *Anatotitan* specimens were identified as large examples of *Edmontosaurus*. *Anatotitan*, however, most likely shared its Late Cretaceous habitat with *Edmontosaurus* and other hadrosaurs. Like these relatives, *Anatotitan* would have been comfortable walking on two legs, but as one

of the Museum's skeletons in the new halls shows, it might also have crouched on all fours to feed, rest, or even walk. *Anatotitan* ate vegetation; its strong jaw muscles and ridged cheek teeth would have enabled it to grind up even the coarsest leaves, needles, and other plant material.

ANKYLOSAURUS

Order: Ornithischia
Suborder: Thyreophora
Family: Ankylosauridae
Late Cretaceous
A skull and tail with plates from different animals discovered in 1910 by Barnum Brown and Peter Kaisen near Red Deer River, Alberta, Canada

"This is without doubt the richest Cretaceous deposit in America," Barnum Brown wrote to Museum curator of vertebrate paleontology Henry Fairfield Osborn at the end of an extraordinary field season in the Canadian badlands along the Red Deer River. "We have up to the present writing filled seventeen boxes of choice material from this formation and have not yet reached the richest field that I examined last year."

Among the many dinosaurs excavated by Brown, Kaisen, and their fossil-hunting team in Alberta during the summers of 1910 and 1911 were the partial skeletons of *Ankylosaurus*, the most famous of the armored dinosaurs commonly called ankylosaurs. One of the finds was a fine tail, called "an extraordinary affair" by Brown, who at first mistook some of the stiff, rigid tail bones as the horn cores crowning the skull of a ceratopsian dinosaur.

The prize specimen, however, was a skull that was first located in 1910. "The *Ankylosaurus* skull is a perfect beauty, uncrushed, all plates in position," Brown wrote Osborn on October 4. The skull was taken up at the end of the field season, and the remaining portions of the associated skeletons were excavated the next year.

THE LIVING ANIMAL

The largest and one of the last of the ankylosaurs, *Ankylosaurus* reached more than thirty feet in length. Like its relatives, it was probably a slow-moving quadrupedal vegetarian. Its back and sides were protected by spines, spikes, and heavy plates of bone; its skull was covered by smaller plates. *Ankylosaurus*' long, muscular tail ended in a large bony club, which it might have used to defend itself from attack by *Tyrannosaurus* or other predators.

Ankylosaurus

APATOSAURUS

Order: Saurischia

Suborder: Sauropodomorpha

Infraorder: Sauropoda

Late Jurassic

Discovered in 1898 by Walter Granger in Como Bluff, Wyoming

Originally called *Brontosaurus*, *Apatosaurus* is rivaled only by *Tyrannosaurus* as the most famous dinosaur skeleton on display in the halls of the Museum. Surprisingly, until the construction of the new halls and the relocation of the dinosaurs to their new home, the huge sauropod had been wearing the wrong head. The story of how *Apatosaurus* ended up with a *Camarasaurus* skull provides a revealing glimpse of how the desire for publicity can overrun the demands for scientific accuracy—even at such an esteemed institution as the Museum.

The Museum's *Apatosaurus* was not the first specimen ever found. That honor belonged to fossils unearthed by Othniel Charles Marsh, who had excavated at Como Bluff from 1879 to 1880. Unfortunately, that skeleton's head was missing—but this setback didn't stop Marsh. His scientific illustration of *Apatosaurus* provided the animal with a short, deep head, modeled on a skull he'd found a remarkable four hundred miles away from where he'd unearthed the rest of the skeleton.

The Museum's specimen, unearthed by paleontologist Walter Granger amid a trove of other sauropod bones at Como Bluff, was also headless (skull bones are far more fragile than the larger bones of the body and often do not get preserved with the rest of the skeleton). As Henry Fairfield Osborn, curator of the Department of Vertebrate Paleontology at the Museum, knew, however, this most spectacular of all dinosaur skeletons would lose much of its impact if mounted without

a skull. He chose to install a model based on Marsh's restoration, although he must have known that it wasn't exactly good scientific practice to assume a skull and skeleton found four hundred miles apart belonged to the same animal.

In 1905, after seven years of restoration, the Museum's *Apatosaurus* went on display, and no one doubted that it was an accurate representation of the long-extinct sauropod. That is, no one expressed any doubt until 1915, when W. J. Holland, director of the Carnegie Museum in Pittsburgh, reported the discovery of a new *Apatosaurus* skeleton. Close by lay

The old, incorrect *Apatosaurus* skull.

a skull, far longer and slimmer than the one chosen by Osborn to crown his specimen.

Although Holland was certain that the American Museum was displaying a fraudulent skull, he never summoned the nerve to install the new, more slender skull on the Carnegie specimen, choosing instead to leave it headless. Such was Osborn's influence at the time.

Despite what must have been overwhelming temptation, Holland never even wrote a refutation of the American Museum's restoration, although he threatened to do so for more than two decades. His only published comment on the matter was a 1915 article in a museum journal. Entitled "Heads and Tails," the piece described how Osborn, "in a bantering mood," had dared him to mount the new head. It was a dare he could not take.

When Holland died in 1935, the Carnegie Museum quickly installed a skull resembling that mounted on the American Museum's specimen. Not until the 1970s, when physicist John McIntosh and paleontologist David Berman went back through the odd history of *Apatosaurus* and stated publicly that the Museum skull was the wrong one, did the public and the scientific establishment alike learn what Holland had suspected sixty years earlier: *Apatosaurus*, now known to be related to the slender-skulled *Diplodocus*, not *Camarasaurus*, deserved a new head.

Now, at last, with the unveiling of the new exhibition halls at the Museum, visitors will see the animal with its proper head.

THE LIVING ANIMAL

While the history of the Museum's *Apatosaurus* is anything but typical, the dinosaur itself was in many ways a typical sauropod. Living during the Late Jurassic period, which was the time of the greatest sauropod diversity, *Apatosaurus* reached about eighty or more feet in length and may have weighed thirty tons. Like other sauropods, it probably moved slowly on four legs, perhaps rearing onto its hind feet to reach treetop leaves and other vegetation or to defend itself against attack by *Allosaurus* or other carnosaurs.

The correct *Apatosaurus*.

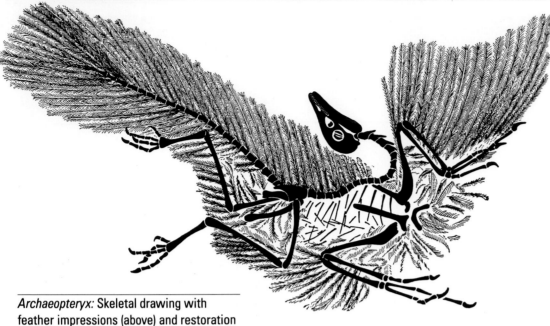

Archaeopteryx: Skeletal drawing with feather impressions (above) and restoration of specimen in same position (below).

ARCHAEOPTERYX

Order: Saurischia
Suborder: Theropoda
Infraorder: Aviale
Late Jurassic
Casts of specimens discovered from 1861 to 1956 in Solnhofen, Germany

Found in the rich limestone deposits of Solnhofen in 1861, the first specimen of *Archaeopteryx* aroused immediate controversy. Only two years earlier, Charles Darwin had published *On the Origin of Species*, and here was a small fossil animal that seemed to support Darwin's theories.

With its reptilian long, bony tail and toothy jaws, and such birdlike features as wing and tail feathers, *Archaeopteryx* seemed to provide an evolutionary missing link between birds and dinosaurs. The strength of the fossil evidence was such that anti-evolution forces sought for decades to prove that the 1861 specimen and others found subsequently were hoaxes.

Even today, *Archaeopteryx* remains controversial, with paleontologists disagreeing on how closely it was related to other early birds, as well as to such birdlike (but, as far as we know, featherless) dinosaurs as *Coelophysis* and *Deinonychus*. Few paleontologists today think that *Archaeopteryx* was actually a direct ancestor of living birds; instead, it may have represented a dead-ended evolutionary pathway. However, most scientists still believe that this small creature supports the theory that birds are, in fact, the only living descendants of the dinosaurs.

THE LIVING ANIMAL

Three-foot-long *Archaeopteryx* was probably not as good of a flyer as most modern birds. Some scientists think it may have used the long claws that tipped its wings to clamber up from the ground to tree limbs or bushes, where it then swooped down on insects or other tiny prey. Its wings may also have allowed it to flutter out of the reach of larger predators.

Coelophysis

COELOPHYSIS
Order: Saurischia
Suborder: Theropoda
Infraorder: Ceratosauria
Late Triassic
Many individuals discovered in 1947 and 1948 by George O. Whitaker, Edwin H. Colbert, Carl Sorenson, and others at Ghost Ranch, New Mexico

One of the earliest dinosaurs that ever existed and one of the first to be described, *Coelophysis*, named for its hollow bones by paleontologist Edward Drinker Cope in 1887, was long known only from frustratingly incomplete fossil fragments. Then, during a 1947 expedition into the Triassic exposures at Ghost Ranch, New Mexico, Museum paleontologist George O. Whitaker followed a trail of fossil scraps up a crumbling hillside to a ledge of bone eroding out of the hill. He brought some samples to fellow Museum fossil hunter Edwin Colbert, who spotted a tiny claw among the bones.

Colbert immediately recognized that the claw belonged to *Coelophysis*. Writing in *Natural History* magazine, he described his emotions when he came to this realization: "As I looked at it, I felt the excitement that comes to one who glimpses treasures in the earth. For years we had hoped to find traces of these primitive little dinosaurs, and the features shown by these fossils could not be mistaken."

Whitaker's initial discovery was just the beginning. The fossil site turned out to be a vast bone bed containing the remains of dozens of *Coelophysis*, including several fully articulated skeletons. Colbert's letters to Museum curator Bobb Schaeffer during the summer of 1947 are filled with the glee of a gold miner tapping into a rich vein: "You ask if we have a skull in our *Coelophysis* material," he wrote on July 11. "To date we have five skulls—one of them a beauty." On July 17, he added, "The more we dig, the more we find."

Among the skeletons unearthed during the summers of 1947 and 1948 were a few specimens whose abdominal cavities contained the bones of young *Coelophysis*. These finds caused Colbert to hypothesize that, perhaps, this dinosaur gave birth to live young instead of laying eggs. But he soon concluded that *Coelophysis'* pelvic opening was too small to permit a live birth; his further research also revealed that the bones came not from embryos but from juvenile *Coelophysis*.

Disproving the live-birth theory left but a single alternative: "The inescapable conclusion is that *Coelophysis* was cannibalistic, eating its own young on occasion, just as do some modern reptiles," Colbert wrote in his memoir, *A Fossil-Hunter's Notebook*. "Not a pretty picture, but a realistic one."

THE LIVING ANIMAL
Whatever natural disaster (perhaps a mud slide) killed the dozens of *Coelophysis* at Ghost Ranch, their numbers convinced some modern scientists that the small, primitive theropod at times traveled in groups. Reaching about ten feet in length, hollow-boned *Coelophysis* weighed only about fifty pounds. With its clawed hands, slender jaw lined with sharp teeth, and powerful hind limbs, it must have been a fast-moving, agile predator. Most likely, *Coelophysis* ate whatever it could catch: small reptiles, mammals, or young of its own kind.

EDMONTOSAURUS

Order: Ornithischia
Suborder: Ornithopoda
Family: Hadrosauridae
Late Cretaceous
Discovered in 1908 by George H. Sternberg in Converse County, Wyoming

During the decades they spent excavating fossils in North America, the dinosaur dynasty of Charles H. Sternberg and his sons, Charles M., George, and Levi, could boast many memorable finds. By their own consensus, however, the Sternbergs were proudest of the *Edmontosaurus* (called *Trachodon* and until just recently called *Anatosaurus*) George discovered in Wyoming. And they had every

right to be proud, for this was the famous "mummy" dinosaur, the first ever found whose skin impressions had fossilized along with the bones.

From the outset, the American Museum's involvement in the 1908 expedition was unexpected. To begin with, the Converse County region had been previously searched by Museum paleontologists with little success, leading to the conclusion that the area did not contain important fossils. Second, Sternberg, a free-lance fossil hunter who had come to search for a *Triceratops* skull for the British Museum of Natural History, had granted his London-based sponsors first rights to any other fossils he might uncover.

The three Sternberg sons had little previous experience searching for dinosaurs, but they made up for their lack of expertise with unquenchable energy and enthusiasm. In

fact, while the elder Sternberg did locate a *Triceratops* skull for the British Museum, it was his son George who found the *Edmontosaurus* bones weathering out of the sandstone. And it was George and his brother Levi who, while excavating the specimen, discovered that many of its bones were wrapped in fossilized impressions of skin, tendons, and even flesh.

When the sons brought their father to the site, Charles H. Sternberg was almost overcome with excitement. "Shall I ever experience such joy as when I stood in the quarry for the first time, and beheld lying in state the most complete skeleton of an extinct animal I have ever seen, after forty years experience as a collector!" he wrote in his 1909 memoir, *Life of a Fossil Hunter*. "The crowning specimen of my life work!"

Despite his prior agreement with the British Museum, Sternberg announced the discovery in a September 2, 1908, letter to Henry

Edmontosaurus

Fairfield Osborn, curator of the American Museum's Department of Vertebrate Paleontology. In the correspondence, Sternberg's tone is professional, almost casual, but it's easy to sense the excitement bubbling just beneath the surface: "It is with great pleasure I write you that we have discovered a *complete* skeleton of *Trachodon*…most of the bones covered with the impressions of the beautiful skin."

Understandably, this announcement caused quite a stir at the American Museum. The British Museum's right of first refusal notwithstanding, Osborn immediately dispatched Albert "Bill" Thomson, one of the Museum's finest collectors, to investigate.

Thomson, however, did not have much success in determining the value of the specimen. By the time the American Museum's emissary arrived in Wyoming, Sternberg had shellacked his find, covered it with flour paste and burlap, and packed it in plaster of paris and wooden crates. In addition, his asking price was two thousand dollars—sight unseen—which was an enormous sum in 1908. "From [Sternberg's] account it must be a very choice thing, but we only have his word for it," Thomson wrote to Osborn in frustration. In a letter to

Museum curator W. D. Matthew, he added, "It strikes me that $2,000 is a stiff price to pay for a pig in a bag."

Despite Thomson's doubts, Osborn had already decided that the purchase of the specimen was essential. And when he heard that his rival, W. J. Holland of the Carnegie Museum in Pittsburgh, had for other purposes also made the trip to Wyoming, actually arriving the same day as Thomson but failing to see Sternberg on the station platform at Lusk, Osborn knew he had to act.

"I telegraphed at once to Sternberg, feeling it was best to secure an option on this specimen," he wrote to Thomson on September 18, adding, "I am glad the Carnegie Museum did not come in on this proposition."

It is unclear how much the American Museum ended up paying for the specimen or how Osborn dealt with Sternberg's prior deal with the British Museum. Typically, however, Osborn's instincts were on target: Since its installation, the fascinating "mummy" *Edmontosaurus* has been one of the most popular and most talked-about displays in all the Museum's exhibition halls.

THE LIVING ANIMAL

The unearthing of the *Edmontosaurus* "mummy" was far more than a rare and fascinating discovery. Indeed, more than any other early-twentieth-century find, it affected our perceptions of how dinosaurs might have looked and acted.

For example, paleontologists learned that the skin of *Edmontosaurus* was not composed of overlapping scales (such as a snake's) but instead was covered with tiny bumps called tubercles. The tubercles were larger in certain areas, indicating that the skin exhibited varied patterns, perhaps even different colors.

Remarkably, the contents of the *Edmontosaurus'* stomach were preserved, along with the skin and other tissue. Inside the stomach were the fossilized remains of such dry-land vegetation as pine needles, bark, and cones, which the hadrosaur had chewed with its thousands of ridged cheek teeth. Previously, most scientists had believed that all hadrosaurs were swamp- and lake-dwelling creatures that feasted on soft aquatic vegetation, but the mummy proved them wrong.

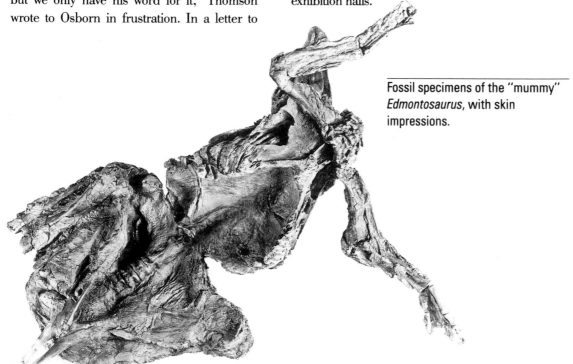

Fossil specimens of the "mummy" *Edmontosaurus*, with skin impressions.

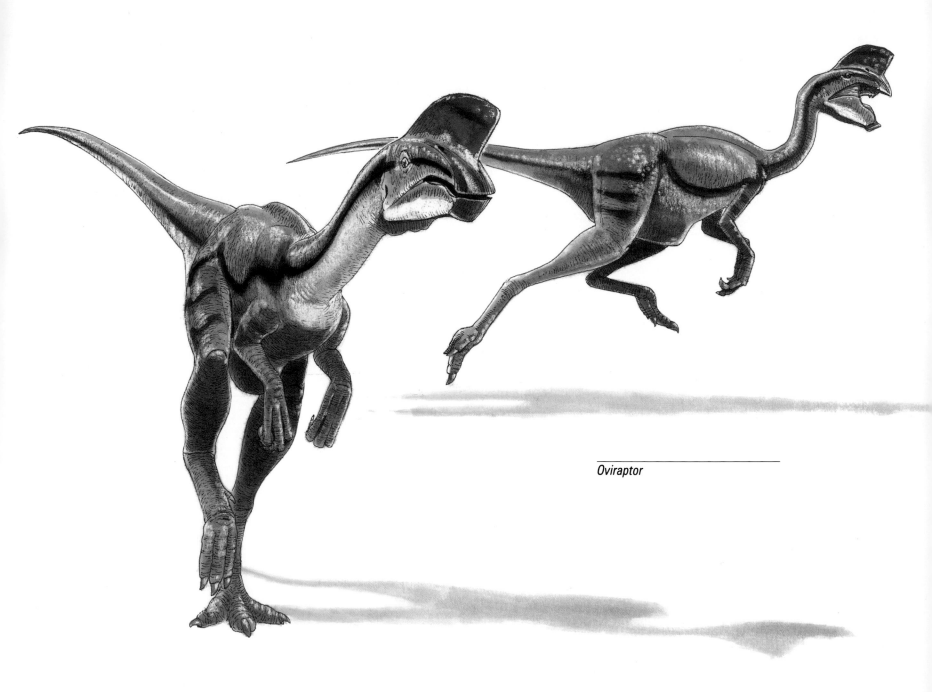

Oviraptor

OVIRAPTOR

Order: Saurischia
Suborder: Theropoda
Infraorder: Coelurosauria
Late Cretaceous
Discovered in 1923 by George Olsen in Shabarakh Usu, Mongolia

During the Museum's 1923 excavation of the Flaming Cliffs in Shabarakh Usu in Mongolia, paleontologists led by Roy Chapman Andrews made a spectacular double discovery. While paleontologist George Olsen was excavating a *Protoceratops* nest, he came upon the partial remains of *Oviraptor*—a bizarre, beaked theropod that lived at the same time as the famous ceratopsian—lying just above the clutch of eggs.

Andrews and chief paleontologist Walter Granger speculated that the little predator had been raiding the nest when it was killed in the same cataclysm that doomed the eggs themselves. "Our fore limb with its long, sharp pointed finger was stretched out as if in the act of scratching out the eggs from their sandy bed," Granger wrote from Peking to Museum curator W. D. Matthew. "Caught in the act without the slightest trace of doubt."

Today, paleontologists aren't so sure what *Oviraptor* ate. But its full name—*Oviraptor philoceratops*, which translates as "egg thief with a taste for ceratopsian eggs"—stands as testament to the incriminating evidence discovered that day in Mongolia.

THE LIVING ANIMAL

Oviraptor reached only about six feet in length. Its strong shoulders, arms, and three-fingered hands would have allowed it to grasp its food. While it may have eaten *Protoceratops* eggs, some paleontologists have guessed that *Oviraptor*'s high, flattened skull, curved jaws, and powerful, virtually toothless beak might have been designed for crushing hard objects —perhaps even bones or mollusks, although no supporting evidence has been found.

PACHYCEPHALOSAURUS

Order: Ornithischia
Suborder: Marginocephalia
Family: Pachycephalosauridae
Late Cretaceous
Discovered in 1940 by William Winkley in Powder Hill, Montana

Sometimes a dinosaur boasts a skeletal feature that seems to serve no obvious purpose. Since paleontologists can't observe the living animal, they must use their knowledge of modern animal physiology and behavior to hypothesize a purpose for these features.

Take *Pachycephalosaurus*. It and other members of its family are named for their most striking characteristic: the remarkably thick, domed roof of the skull, which in *Pachycephalosaurus* was a full ten inches thick.

Despite its impressive thickness, *Pachycephalosaurus*' skull probably did not aid directly in an individual animal's survival. Since the rest of the dinosaur's body was unprotected, the dome wouldn't have provided much deterrence to predators nor could it have served as an aid in the search for food.

Today, many paleontologists believe that *Pachycephalosaurus* may have used its bony dome in ritual battles for dominance over a herd or family group. Much as bighorn sheep clash horns today, the dinosaurs may have butted heads yet suffered little or no injury due to the specialized domes and a specially strengthened backbone. However, as is the case with so many of the theories about dinosaur behavior, this will never be more than an educated guess.

THE LIVING ANIMAL

Although this fifteen-foot-long, largely bipedal dinosaur resembled an ornithopod in many ways, paleontologists now believe that *Pachycephalosaurus* was more closely related to the ceratopsians, the group that included such famous dinosaurs as *Triceratops*. Most likely a plant eater, it may have used its upright posture to reach vegetation beyond the reach of the resolutely quadrupedal ceratopsians.

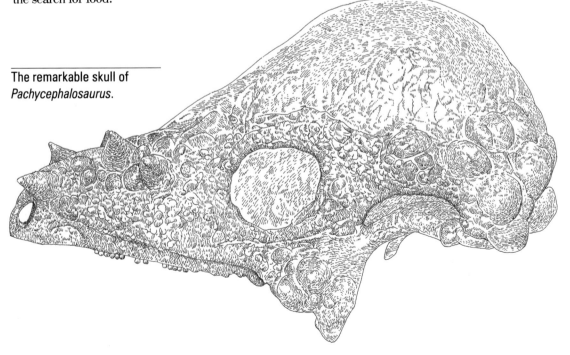

The remarkable skull of
Pachycephalosaurus.

PENTACERATOPS

Order: Ornithischia
Suborder: Marginocephalia
Family: Ceratopsidae
Late Cretaceous
Discovered in 1922 by Charles H. Sternberg near Tsaya, New Mexico

Without a doubt, Henry Fairfield Osborn, curator of the Museum's Department of Vertebrate Paleontology, would have loved to have stocked the Museum's dinosaur collection entirely with fossils found by his own staff. When star fossil hunters Barnum Brown or Walter Granger, for example, reported a discovery, Osborn could be certain that he was receiving an accurate description of the find and that the specimen automatically belonged to the Museum.

When the collector was free-lance bone digger Charles H. Sternberg, however, the road was usually bumpier. Sternberg, an emotional man, sometimes overestimated the value of his finds and could be (at least in Osborn's view) unreasonable in his demands. In spite of these quirks, Sternberg did have a tremendous ability to locate important specimens, and Osborn knew it.

The negotiations surrounding the sale of the skull of *Pentaceratops* were typically arduous. Sternberg thought he'd found the remains of a *Triceratops*, but his description of the specimen's skull made it clear to Osborn and curator W. D. Matthew that he'd actually found something much more spectacular. "The '*Triceratops*' is an extraordinary thing," Matthew told Osborn. "I strongly recommend that even at a considerable price you should secure this specimen."

This course of action, however, was easier to recommend than to accomplish. The two parties were eventually able to reach an agreement, and as usual, Osborn got what he wanted. In the end, Matthew was correct: The animal turned

Pentaceratops, and the skull unearthed by Charles H. Sternberg—the first *Pentaceratops* specimen ever found.

out to represent a new genus, *Pentaceratops sternbergii*. Today, the enormous skull, mounted on a plaque, remains one of the most bizarre and spectacular fossils on display at the American Museum.

THE LIVING ANIMAL

Matthew named Sternberg's find *Pentaceratops* in honor of its five horns: two located on the brow, another on the nose, and two more, far shorter and duller, adorning the cheeks. Perhaps the most remarkable feature of *Pentaceratops*, however, was its huge neck frill; the head and frill together measured more than seven feet in length and made up a third of the dinosaur's overall length.

Paleontologists remain divided on the use of the showy frills of *Pentaceratops* and other ceratopsians. Some believe that the frills served as shields to defend the dinosaurs' vulnerable backs from attack. Others disagree, pointing out that the bones of the frills were fragile and probably useless for defense and maintaining that *Pentaceratops*' frill, with its large surface area, may have served to anchor the strong muscles that supported the dinosaur's massive head and jaws.

PLATEOSAURUS

Order: Saurischia
Suborder: Sauropodomorpha
Infraorder: Prosauropoda
Late Triassic
Discovered in 1925 by Friedrich von Huene in Trössingen, Germany

Europe, the birthplace of paleontology, was home to many fascinating dinosaurs and other ancient creatures. The famous dinosaur-bird *Archaeopteryx* and the hen-size theropod *Compsognathus* roamed what is now Germany, while England played host to the hulking ornithopod *Iguanodon* and the bizarre, meat-eating *Baryonyx*, with its slashing claw and crocodilelike jaws. Perhaps the dinosaur that was most widespread in Europe was the prosauropod *Plateosaurus*, whose fossils have turned up in many sites in France, Switzerland, and Germany. The specimen on display in the American Museum's halls is one of a pair unearthed by the great German paleontologist Friedrich von Huene, who undertook a special expedition for the American Museum in 1925.

Named in 1841, *Plateosaurus* was also one of the first dinosaurs to be identified. Along with *Iguanodon*, *Megalosaurus*, *Hylaeosaurus*, and others (many of which have either been renamed or discredited as legitimate genera), the discovery of *Plateosaurus* fossils helped spur the dinosaur gold rush that swept across the continents of Europe and North America in the mid- to late 1800s.

THE LIVING ANIMAL

Twenty-six-foot-long *Plateosaurus* was among the largest of the prosauropods. It resembled the later sauropods, such as *Apatosaurus*, in many ways—including its thick body, pillar-like hind limbs, and long, slender neck and tail. Although paleontologists think that *Plateosaurus* and other prosauropods were only distantly related to the sauropods, most believe that the two groups evolved from a common ancestor.

Plateosaurus

During the American Museum's famous series of expeditions to Central Asia between 1919 and 1930, Roy Chapman Andrews and his team of scientist-adventurers made many astonishing finds. The great fossil bed at the Flaming Cliffs of Shabarakh Usu in Mongolia, however, may have been Andrews' crowning achievement.

The richness of the Flaming Cliffs site was discovered at the tail end of the 1922 expedition, during the team's journey home. Returning to the site on July 8 of the following year, the explorers found that the Flaming Cliffs fossil beds surpassed their wildest imaginings. Within the first two days of digging, they had unearthed nearly a dozen *Protoceratops* skulls, two still boasting the fragile neck frills. During the next five weeks, the team excavated dozens of skeletons, including remains of baby, juvenile, and adult *Protoceratops*. "The most important discoveries, however, proved to be two nests of dinosaur eggs," Andrews wrote coolly in his journal.

"The great series of skulls will make a splendid exhibit," Andrews predicted in a 1923 letter to Museum curator W. D. Matthew. Today, the skulls, skeletons, and eggs remain among the proudest possessions in the Museum's entire dinosaur collection.

PROTOCERATOPS

Order: Ornithischia
Suborder: Marginocephalia
Family: Protoceratopsidae
Late Cretaceous
Many skeletons and eggs, most discovered in 1923 by Walter Granger, Peter Kaisen, George Olsen, and others in Shabarakh Usu, Mongolia

THIS PAGE, FROM TOP TO BOTTOM: Protoceratops; Protoceratops skull and bones before being removed from Mongolian sand and rock; the famed Protoceratops eggs—the first dinosaur eggs to be identified.

THE LIVING ANIMAL

Protoceratops was one of the earliest and most primitive of the ceratopsians. It was small, with adults only reaching about seven feet in length. Unlike later relatives, it had no horns, only a thickened mass of bone on its nose and small bumps above its eyes. And while *Protoceratops* probably was quadrupedal like all later ceratopsians, its hind legs were far longer than its forelimbs. This feature, along with its frill, demonstrates its kinship with such primitive ceratopsians as *Psittacosaurus*, which walked on two legs.

STEGOSAURUS

Order: Ornithischia
Suborder: Thyreophora
Family: Stegosauridae
Late Jurassic
Discovered in 1901 by Peter Kaisen in Bone Cabin Quarry, Wyoming

No fossil site on United States soil was more productive than the Late Jurassic beds around Como Bluff, Wyoming; on his first visit, Barnum Brown, the Museum's star fossil hunter, reported seeing a myriad of bones simply lying on top of the ground. It was here that the legendary paleontologist Othniel Charles Marsh made some of his finest discoveries. And it was here in 1898 that Museum curator Walter Granger discovered the Museum's famous (and, until recently, wrong-headed) *Brontosaurus* (now renamed *Apatosaurus*).

The Como Bluff region was so rich that it took years to work through the area's fossil beds. During that time, Museum and other paleontologists found the bones of *Apatosaurus*, *Diplodocus*, *Camarasaurus*, and many others, including the enormous *Stegosaurus* skeleton that graces the Museum's halls.

THE LIVING ANIMAL

Stegosaurus—the only plated dinosaur found in North America to this day—was, at twenty-five feet or more in length, one of the largest of its family. It is believed to have walked on all fours, plucking vegetation with its toothless, beaklike mouth and chewing with the leaf-shaped teeth that lined the back of its jaws.

Scientists continue to debate the purpose of the large bony plates that lined *Stegosaurus*' back and upper tail. Perhaps they served as a defense, causing large predators to choose more vulnerable prey. Or they may have functioned as heat exchangers, providing extra skin area to soak up the sun's warmth in cool weather and vent excess heat during hot spells.

Stegosaurus

TENONTOSAURUS

Order: Ornithischia
Suborder: Ornithopoda
Family: Iguanodontidae
Early Cretaceous
Discovered in 1932 by Barnum Brown and Peter Kaisen in Mott Creek, Montana

The remarkable Early Cretaceous ornithopod *Tenontosaurus* was discovered on a Crow Indian reservation by Museum fossil hunters Barnum Brown and Peter Kaisen in 1932. Mounted for display in 1938, the skeleton remained unidentified. Brown, Kaisen, and their fellow paleontologists decided that it most closely resembled a camptosaur (a primitive relative of the famous *Iguanodon*), but in truth, *Tenontosaurus*, with its flexible neck and extraordinarily long, heavy tail, really didn't look like any other dinosaur at all.

The uncertainty continues to this day. Some experts classify twenty-four-foot-long *Tenontosaurus* with the hypsilophodontids, but this seems an awkward fit, for all other known hypsilophodontids were far smaller and slimmer than the bulky *Tenontosaurus*. The paleontologists at the American Museum have placed this unique ornithopod back where Barnum Brown thought it belonged—with the heavy-bodied, slow-moving iguanodontids.

THE LIVING ANIMAL

Even if it was most closely related to *Iguanodon* and similar ornithopods, *Tenontosaurus* boasted many characteristics not seen in those largely bipedal plant eaters. For example, many paleontologists believe that it spent much of its time on all fours, perhaps rising occasionally onto its hind limbs to reach particularly desirable vegetation. Even more unusual was *Tenontosaurus'* tail, which took up more than half of the dinosaur's overall length.

A fossil bed in Montana contained both a *Tenontosaurus* skeleton and five skeletons of *Deinonychus*, the magnificent sickle-clawed theropod. Although we'll never know for sure how these fossils ended up in the same location, this find allows us to imagine a pack of thirteen-foot-long *Deinonychus* chasing down and killing the much larger ornithopod, much as African hunting dogs cooperate to kill a zebra or antelope far too large to be challenged one-on-one.

Tenontosaurus

TRICERATOPS

Order: Ornithischia
Suborder: Marginocephalia
Family: Ceratopsidae
Late Cretaceous
Most material for the composite skeleton discovered in 1909 by Charles H. Sternberg, Barnum Brown, and Peter Kaisen in Wyoming and Montana

During the great nineteenth-century bone war waged by fossil hunters Edward Drinker Cope and Othniel Charles Marsh, both paleontologists found abundant remains of *Triceratops*. Cope's men were the first to discover the great ceratopsian, but the fossils were too poor in condition to permit conclusive identification, and it was left to their rival, Marsh, to give *Triceratops* its name.

Marsh, however, also had trouble determining the proper identification for this spectacular dinosaur; on digging up the first well-preserved skull, he thought he'd found the remains of a gigantic bison.

The Museum's skeleton comprises bones from several individuals, which suggests *Triceratops*' apparent abundance in North America about sixty-five million years ago. The field journals and correspondence of many digs in the Late Cretaceous sites of Wyoming, Montana, and Alberta show that Barnum Brown and other Museum paleontologists frequently found remains of *Triceratops*, although skulls were scarce and most material was poorly preserved. As was the case with *Edmontosaurus*, the fossil hunters often seemed to be looking right past *Triceratops* bones, hoping for something more exciting under the next layer of earth or rock.

THE LIVING ANIMAL

Fossil evidence, such as the sheer abundance of *Triceratops* bones in Late Cretaceous beds, has led many paleontologists to postulate that the thirty-foot-long ceratopsians traveled in herds. *Triceratops* may have behaved much as bison did in more recent times, following a seasonal route through the vast plains of western North America in a constant search for food.

Despite many famous restorations, little evidence exists that *Triceratops* was hunted by *Tyrannosaurus* or other carnosaurs. It is possible, however, that North American theropods did follow the herds, looking for the opportunity to fell young or sick *Triceratops*—or perhaps even a healthy adult.

Triceratops.

TYRANNOSAURUS REX

Order: Saurischia
Suborder: Theropoda
Infraorder: Carnosauria
Late Cretaceous
Discovered in 1908 by Barnum Brown in Hell Creek, Montana

By the time Museum fossil hunter Barnum Brown set up camp in the badlands of Montana in the summer of 1908, the area was already well known as a rich site for Late Cretaceous fossils. Local ranchers had long reported finding fossilized bones in the area, and only six years earlier, Brown himself had made a spectacular discovery nearby: the first largely complete *Tyrannosaurus* skeleton ever unearthed.

Fossils, including many of the duck-billed dinosaur *Edmontosaurus*, were plentiful. However, Brown's letters to Henry Fairfield Osborn, chief curator of the Museum's Department of Vertebrate Paleontology, showed that the fossil hunter was expecting even more from the 1908 field season.

Early in July, Brown's greatest hopes were realized. On July 8, he wrote to Osborn, "At last I have some good news to report. I made a ten-strike last week," finding the extensive remains of what was clearly a huge—and well-preserved—carnivorous dinosaur.

At first, Brown didn't know exactly what it was he'd found. "This is another new dinosaur," he wrote on July 8. "I have seen nothing like it." By July 15, however, his confusion had vanished, and he sent word to Osborn: "Our new animal turns out to be *Tyrannosaurus*."

In all of Osborn's correspondence at the Museum, no letter resounds with the joy that fills the reply he sent to Brown on July 30. "Your letter of July 15 makes me feel like a prophet and the son of a prophet, as I felt instinctively

---

OPPOSITE: Tyrannosaurus rex menacing Triceratops. RIGHT: Tyrannosaurus head.

that you would surely find a *Tyrannosaurus* this season," he wrote. "I congratulate you with all my heart on this splendid discovery."

The enormous skeleton (the block of stone containing the skull alone weighed fifteen tons) was shipped back to New York City and restored, and for many years, the Museum was home to the only two reasonably complete *T. rex* skeletons on earth. During World War II, however, fears arose that the city might be bombed, sending the Museum's dinosaurs into a second mass extinction. While the 1908 specimen was retained, the 1902 skeleton was promptly sent to the Carnegie Museum of Natural History in Pittsburgh, where it still resides today.

THE LIVING ANIMAL

In its original exhibited posture, with its head held high and tail dragging along the ground, the 1908 *T. rex* remains one of the most familiar and impressive dinosaur exhibits on earth. It is unlikely, however, that the great meat-eater actually maintained such an upright position,

at least while moving. Most scientists now believe that *T. rex*, like other dinosaurs, usually stood with its back held horizontally, almost parallel to the ground. Its tail, stiffened by strong tendons, would have stretched into the air behind the body, serving as a counterweight as the dinosaur strode forward. The posture of the newly restored skeleton in the Museum's new exhibition hall reflects this current thinking.

Whatever its posture, forty-foot-long *T. rex*, with its huge head, which was five feet in length, and knife-edged, seven-inch teeth, was clearly one of the largest and most powerful carnivores in the history of life on earth. Remarkably, some scientists believe that *T. rex* might actually have been *too* large to have been an efficient predator and instead scavenged the bodies of animals that were already dead. Most experts, however, still envision *Tyrannosaurus rex* as a hunter that utilized a combination of size, speed, and power to chase down and kill its prey.

GALLERY OF MAMMALS

ANDREWSARCHUS
Order: Arctocyonia
Family: Mesonychidae
Late Eocene
Discovered in 1923 by Kan Chuen Pao, known as "Buckshot," in Irden Mannah, Mongolia

The day the cars returned with mail, the Third Asiatic Expedition held a celebration—mail from home, and that same afternoon Buck Shot, one of our remarkable Chinese assistants, working with Olsen, discovered a perfect skull of Entelodon—*or what corresponds to that genus.*

This passage, part of a characteristically ebullient letter from Roy Chapman Andrews to the Museum's Department of Vertebrate Paleontology curator Henry Fairfield Osborn on June 8, 1923, contains the first announcement of the discovery of *Andrewsarchus*—a strange-looking mammal still known chiefly from that single, perfect three-foot-long skull. Despite his enthusiasm, Andrews' seeming certainty about the discovery ("*Entelodon*—or what corresponds to that genus") actually masked a great deal of doubt and foreshadowed a debate over *Andrewsarchus'* habits that still sputters on today.

A look at Andrews' unpublished field journals indicate that at first he identified the find as the "great carnivore *Mesonyx.*" Within a day or two, however, chief paleontologist Walter Granger convinced him to change the identification to the "artiodactyl *Entelodon*"—describing not a huge carnivore but a hoofed, piglike animal.

Back at the Museum, curator W. D. Matthew studied a sketch of the skull and quickly saw that Andrews' initial identification had been far closer to the truth than Granger's. Matthew thought that the skull clearly belonged to a carnivore that was most likely a member of a primitive group called the creodonts, within the family *Mesonychidae.*

Hearing that his identification had been overruled, Walter Granger was abashed. "The blunder on that Mesonychid was mine," he admitted in a letter to Matthew. "A Hell of a paleontologist who can't tell a (?) creodont from an Artiodactyl."

THE LIVING ANIMAL

"Almost every previous restoration, every previous description, of this animal has been incorrect," declares current Museum curator Malcolm McKenna, himself the veteran of several expeditions to Mongolia. "*Andrewsarchus* is usually portrayed as a fierce carnivore, but its teeth were blunt to start with, and they wore down to mere stumps. I suspect it was living on carrion or bones or mollusks—who knows? But it wasn't slicing up meat."

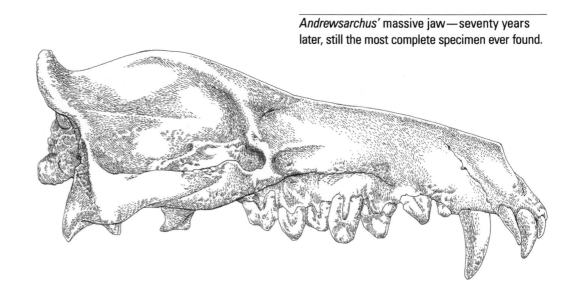

Andrewsarchus' massive jaw—seventy years later, still the most complete specimen ever found.

BRONTOPS
Order: Perissodactyla
Suborder: Hippomorpha
Family: Brontotheriidae
Early Oligocene
Collected in 1892 in White River, South Dakota

Brontops skeleton.

Brontotheres, massive plant-eating mammals closely related to horses, were abundant across Oligocene North America. Henry Fairfield Osborn, curator of the Museum's Department of Vertebrate Paleontology, was fascinated by the hard-to-explain variations in the brontotheres (then called the titanotheres), especially because they represented a window into his unusual theory of evolution.

Osborn was particularly intrigued by the brontotheres' "horns," odd outgrowths of the nasal bones that were lacking in primitive genera but, in later members of the family, became

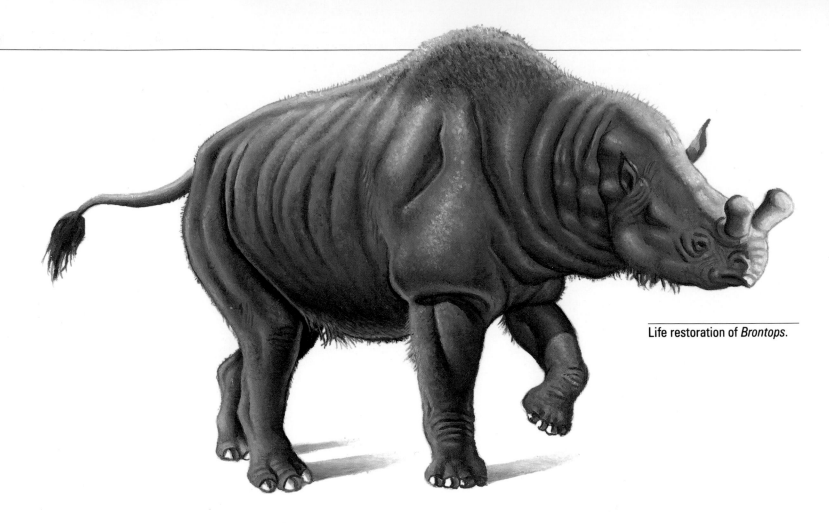

Life restoration of *Brontops*.

huge and ornate. The paleontologist saw these structures not as progressive adaptations to an ever-evolving environment (the Darwinian view), but as an inevitable process of change determined by the animals' genes.

In Osborn's view, these irreversible changes might interfere with a species' ability to survive. Pointing to the extinction of all the brontotheres by the end of the Oligocene, he concluded that their inborn genetic propensities had, in fact, led to their demise.

Osborn's preoccupation with the brontotheres led him to guard the Museum's specimens with jealousy and suspicion. He did not allow other scientists to study and describe them—and only he determined which fossil groups constituted a new genus or species. Unfortunately, this great paleontologist's enthusiasm often overshadowed his scientific

judgment as he eagerly "discovered" one species after another.

Not until 1945, more than a decade after Osborn's death, did a Museum paleontologist publicly discount much of Osborn's deeply held beliefs. In his "A Classification of the Mammals," George Gaylord Simpson wrote, "It is fairly certain that a number of the genera, and quite possible that some subfamilies, have been based on variations that were really sexual, ontogenetic, or individual." Simpson continued with a delicate acknowledgment of Osborn's lapses: "Of course no two individuals are alike, and subjective criteria have seemed to warrant placing virtually every good specimen in a new species."

THE LIVING ANIMAL

Like many brontotheres, *Brontops*, which stood eight feet high at the shoulder, boasted an odd pair of "horns" set near the tip of its snout. Most likely covered in skin, these structures were actually outgrowths of the nasal bones and differed both from the true horns of antelopes and from rhinoceros "horns," which are composed of compacted hair.

Horns and antlers are generally used by animals for species recognition, defense from predators, weapons used in intraspecies battles for dominance over the herd or family group, or some combination of these. Extinct horned and antlered animals—including *Brontops*—often baffle paleontologists. Left with ancient, long-dead clues, scientists can only make educated guesses about what the brontothere really used its horns for.

CANIS DIRUS

Order: Carnivora

Family: Canidae

Pleistocene to Recent

Discovered in Rancho La Brea, California; acquired in 1914

"I am just back from a trip to the La Brea asphalt fossil deposit, which is a truly Amazing Aggregation of Antique Animals and makes all other fossil deposits look sick by comparison," wrote the usually staid Museum curator W. D. Matthew in a 1913 letter. His sense of wonder, almost of disbelief, has been echoed by countless others on their first sight of the extraordinary finds excavated from the seemingly bottomless tar pools of La Brea.

Overall, paleontologists have found remains of more than two thousand *Canus dirus* in the asphalt deposits. Alongside were equally abundant fossils of sabertooth cats and ancient vultures, as well as smaller numbers of mammoths, mastodons, ground sloths, horses, and other animals and birds.

As Matthew pointed out in a 1913 issue of the *American Museum Journal,* the extraordinary number of meat-eating creatures retrieved from the pools paints a vivid picture of events there during the Pleistocene era. "[T]he larger quadrupeds, venturing out upon the seemingly solid

La Brea scene with *Canis dirus* (left and foreground), *Homotherium* (foreground left), *Glossotherium* (center, background center, and background right), *Mammuthus* (background left), and Clovis Indians (background right).

surface and caught in the asphalt, served as a bait for animals and birds of prey, luring them from all the country round about and enticing them within the treacherous clutch of the trap. These in their turn, falling victims, served to attract others of their kind. And so the 'death-

trap of the ages'...self-baiting, automatically disposing of its prey, has collected and preserved to our time a truly wonderful series of the predacious animals and birds."

THE LIVING ANIMAL

A close relative of dogs, wolves, and coyotes, *Canis dirus* resembled these modern canids in shape and habits. More heavily built than a wolf, it may have been omnivorous, eating nuts and berries and scavenging dead meat—except when it came across (seemingly) easy prey, such a horse trapped in the La Brea tar pools.

Scientists think that *Canis dirus,* like most modern canids, may have hunted in packs. Its competitors for food would have included the sabertooth cat *Smilodon* and *Panthera leo atrox,* an extinct subspecies of the same species that includes the modern lion.

Canis dirus skeleton.

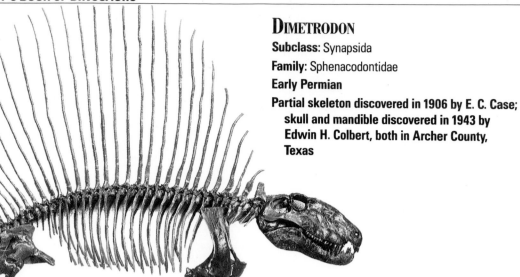

DIMETRODON

Subclass: Synapsida

Family: Sphenacodontidae

Early Permian

Partial skeleton discovered in 1906 by E. C. Case; skull and mandible discovered in 1943 by Edwin H. Colbert, both in Archer County, Texas

THIS PAGE, FROM TOP TO BOTTOM: Dimetrodon skeleton, skull and jaws, and restoration. Although it most closely resembles some great reptile, *Dimetrodon* and other pelycosaurs belong to the line that eventually led to modern mammals.

THIS PAGE, CLOCKWISE FROM TOP: Dimetrodon skeleton, skull and jaws, and restoration. Although it most closely resembles some great reptile, *Dimetrodon* and other pelycosaurs belong to the line that eventually led to modern mammals.

The 1906 specimen, including a fine skull, of the great sail-backed "pelycosaur" (scientists are currently reclassifying *Dimetrodon* and its relatives) *Dimetrodon* was one in the remarkable trove of Permian animals collected by E. C. Case, a fossil hunter and college professor who was on loan to the Museum that summer. His Texas finds and others in the southwestern United States provided scientists with an unparalleled picture of life in the Permian, the period when the "pelycosaurs" were most abundant.

During the Permian, as many as 70 percent of all four-legged terrestrial animals were "pelycosaurs," including such widely varied creatures as the huge *Cotylorhynchus* with its ridiculously tiny head, the carnivorous *Varanosaurus*, and *Dimetrodon*, whose great size, powerful jaws, and enormous canine and shearing teeth confirm that it was a dominant predator of its time.

THE LIVING ANIMAL

Ten-foot-long *Dimetrodon*'s most striking skeletal feature by far was the sail that ran along its back to the base of its tail. Many scientists believe that the sail—made up of elongated spines of the vertebra and linked by a sheet of skin rich in blood vessels—probably served as a "heat exchanger" for this cold-blooded animal. During the cool of the morning, *Dimetrodon* may have turned the sail toward the sun, soaking up warmth and con-

verting it to energy, while perhaps also snapping up sail-less creatures still sluggish from the cold. As the day heated up, some scientists speculate that *Dimetrodon* could have turned the sail away from the sun, allowing the blood vessels to radiate excess heat.

Dimetrodon is not the only ancient animal to bear a sail. Another "pelycosaur," *Edaphosaurus*, boasted a similar sail, as may have at least two unusual dinosaurs: the carnosaur *Spinosaurus* and the iguanodont *Ouranosaurus*.

ENALIARCTOS

Order: Carnivora
Suborder: Pinnipedia
Family: Enaliarctidae
Early Miocene
Discovered in Kern County, California; cast acquired in 1975

Perhaps 220 million years ago, as the dinosaurs were beginning their unparalleled 160-million-year reign on earth, the earliest mammals appeared. For the rest of the age of the dinosaurs, few mammals grew larger than the size of a shrew, and as a group, they apparently did little to expand their range or diversity.

Not until the large dinosaurs became extinct sixty-five million years ago did mammalian evolution enter a new phase. Over the course of approximately ten million years, ruminants

and other plant eaters diversified greatly, feasting on the abundant vegetation, and a myriad of new carnivores—including the ancestors of today's bears, cats, and dogs—also evolved.

Among these carnivores was a group of mammals that returned to the sea, apparently taking advantage of the ecological niches left by the now extinct oceangoing reptiles, the plesiosaurs and pliosaurs. Some of these marine mammals eventually became the first pinnipeds, ancestors of today's seals, sea lions, and walruses. One of these early pinnipeds was *Enaliarctos*.

Scientists believe that the shape of its skull and other skeletal features identify *Enaliarctos* as a descendant of primitive bears. Other characteristics, including feet modified into paddles, large eyes, and a well-developed upper lip and whiskers point to *Enaliarctos* as an early relative of today's pinnipeds.

THE LIVING ANIMAL

Enaliarctos' mix of physical characteristics was probably best suited to life in coastal waters. Like modern sea otters, it foraged in the protected bays off the coast of California. Its diet most likely included a mixture of fish and shellfish.

No pinniped, ancient or modern, ever became as superbly adapted to marine life as the whales and dolphins, who are able to give birth at sea. *Enaliarctos*, like other pinnipeds, probably had to return to land to breed.

Enaliarctos

EQUUS SCOTTI

Order: Perissodactyla

Family: Equidae

Late Pleistocene

Discovered in 1899 by J. W. Gidley in Briscoe County, Texas

I think this about answers the description I heard someone give once, of what they thought a 'bone-diggers' heaven would be, a bone bed of skeletons, just disarticulated enough to make it interesting, in a bed of sand so loose that you could dig it away with your hands, and the bones good and hard.

This bone digger in heaven was J. W. Gidley, one of the American Museum's premier hunters of ancient mammals early this century. The site in question was a quarry of *Equus scotti* bones lying in an extensive Pleistocene fossil bed at Rock Creek, Texas. Clearly, Gidley knew he'd found something special.

Gidley also knew how much his find would please Henry Fairfield Osborn. Throughout his decades as the Museum's curator of vertebrate paleontology, Osborn treasured fossils of horses above almost any other kind. To him and other Museum curators, these discoveries represented more than simply new pieces in the fossil puzzle: They were links in an elegant chain tracing the evolution of horses from the terrier-sized multitoed *Hyracotherium* to the familiar hoofed *Equus*—a chain, Osborn believed, that proved the truth of evolutionary theory beyond all doubt.

ABOVE RIGHT: Charles R. Knight's *Equus scotti.   RIGHT:* A Museum *Equus scotti* skull.

THE LIVING ANIMAL

It is likely that the Late Pleistocene *Equus scotti* closely resembled the modern horse. Remarkably, although ancestral horses in North America experienced a long period of unparalleled abundance that stretched from Eocene to Recent times, all North and South American horses died out about eight thousand years ago and did not return until reintroduction by sixteenth-century European settlers. No one knows for certain why the equine line came to such a sudden and complete end in this hemisphere.

EURHINODELPHIS

Order: Cetacea
Suborder: Odontoceti
Family: Rhabdosteididae
Mid to Late Miocene
Discovered in 1972 in Calvert County, Maryland; cast of specimen acquired in 1972

porpoises. For example, the complex structure of its ear indicates that it might have been able to identify prey in its dim habitat by emitting sounds and then analyzing the sound waves as they bounced back. This ability to echolocate exists in modern dolphins, porpoises, and toothed whales, but few other animals.

Eurhinodelphis' other distinctive feature was its long, swordlike jaws. Lined for much of their length with sharp teeth but toothless at the tip, these jaws probably evolved as yet another weapon against *Eurhinodelphis'* preferred prey. Most likely it could wield its sword with quick, dextrous movements, perhaps stunning fish by slashing at them and then grabbing them with its teeth.

Remarkably, whales and dolphins, the mammals most exquisitely adapted to life in the oceans, seem to have evolved from early ungulates—land-based, largely plant-eating animals that were the forerunners of today's horses, cattle, and deer. Although the fossil record is spotty, the ungulates' return to the marine environment appears to have taken place about forty-five million years ago, as mammals began filling the niches left after the massive marine reptiles, the plesiosaurs and pliosaurs, became extinct.

Eurhinodelphis, which roamed the Pacific coastal waters of North America and Asia about fifteen million years ago, had already evolved many of the features found in modern

THE LIVING ANIMAL

Reaching about six and a half feet in length, *Eurhinodelphis* must have been an active hunter of fish. Its well-developed paddles, muscular tail, and streamlined body would have given it both speed and agility. Add the ability to echolocate, and *Eurhinodelphis* would have been one of the premier predators of its coastal habitat.

Eurhinodelphis

MAMMUT

Order: Proboscidea
Suborder: Elephantoidea
Family: Mammutidae
Late Miocene to Late Pleistocene
Discovered in 1845 in Orange County, New York; acquired in 1906

From its discovery on a New York State farm through its heralded arrival in New York City in 1906 and its much-anticipated unveiling at the American Museum in 1907, the Warren Mastodon was one of the most famous ancient animals on earth.

The great mastodon *Mammut*'s magnificent tusks.

Harvard anatomist John Collier Warren, who acquired the nearly perfect skeleton a few months after its excavation, called attention to his purchase by building the mastodon its own museum in Boston. In 1852, Warren, clearly enamored of his find, published a gigantic tome, *Description of a Skeleton of the Mastodon Giganteus of North America*, dedicated to the beast and complete with stunning illustrations.

The superb specimen's fame was ensured in 1906, when renowned financier J. P. Morgan bought Warren's collection—including the mastodon—and donated the lot to the American Museum. Curator of vertebrate paleontology Henry Fairfield Osborn was, of course, extraordinarily grateful—but not too grateful to consider approaching the Morgan family with a request for even greater largess.

Osborn, who never let modesty get in the way of the Museum's best interests, nevertheless balked at dunning Morgan himself quite so soon after the generous bequest. Instead, he turned to the financier's son, J. P. Morgan, Jr. In an April 1906 letter addressed to "My dear Jack," Osborn wrote with uncharacteristic hesitancy, "It is an old motto, 'not to look a gift horse in the mouth,' but I may break the rule so far as to ask whether you would enjoy helping me work this Warren collection up." He added, "If this does not strike you favorably please promptly and emphatically say no."

Morgan Junior seemed to have no trouble saying no, conveying his sentiments in a letter by return mail. He did, however, offer to pay five hundred dollars a year for the restoration effort—a generous sum at the time and likely all that Osborn was really hoping for.

THE LIVING ANIMAL

When scientists unearthed the bones of a mastodon during construction of an Ohio golf course in 1991, they thought they'd merely found an unusually complete skeleton of *Mammut*, which roamed North America until about ten thousand years ago.

After more extensive searching in the dense, airless earth around the fossils, the paleontologists found something far more exciting: a mass of organic material that they determined to be the remains of the mastodon's last meal. Amazingly, the stomach con-tents included living bacteria; at least eleven thousand years old, they are by far the oldest living organisms ever found.

Until this time, most experts had believed that the mastodons thrived by eating needles and twigs from the spruce forests that covered most of the temperate regions during the Pleis-tocene era. As shown in the Ohio find, however, that individual, at least, had been feeding on soft vegetation that is characteristic of ponds and marshes, a discovery that definitely calls into question previous assumptions of the mastodon's habits.

ABOVE: Charles R. Knight's restoration of the Warren Mastodon.

Charles R. Knight's *Mammuthus*, the mammoth.

MAMMUTHUS

Order: Proboscidea

Suborder: Elephantoidea

Family: Elephantidae

Late Pleistocene

Discovered in Jonesboro, Indiana; acquired in 1904

Dear Sir's.
Find enclosed photoes of the Skull and bones of the Mamoth which we unearthed recently while we were ditching....Kindly let me know if you are interested in anything like this and what it is Worth to you.

This letter, dated May 25, 1909, from a farmer in Ashley, Indiana, is a typical example of the hundreds received by Museum paleontologists during the first third of this century. It seemed you could barely begin plowing a field or digging a ditch without stumbling on the fossilized remains of *Mammut*, the mastodon, or *Mammuthus*, the mammoth—the two great elephants that abundantly roamed the continent during the Late Pleistocene era.

Every so often, however, a find on someone's land was a spectacular one: the skeleton fully articulated, the tusks perfectly preserved. Such

was the case with the specimen now on display in the Museum's new halls. Found on a farm belonging to Dora E. Gift, it is one of the finest mammoth skeletons on display in the world.

What was most striking about most other amateur discoveries, however, was not the finds themselves, which usually included only fragmentary material (often just a single tooth) that was almost always badly weathered or broken up, but the enthusiasm of these incidental fossil hunters and the eagerness with which many of them offered their specimens to the Museum for free.

Some, however, thought their finds would make them rich; at least one would-be fossil hunter demanded $100,000 for his partial mammoth skeleton, only to receive a stern rebuff from a Museum curator.

The Ashley, Indiana, farmer was one who expected to be paid for his find. Discovering that the specimen in question was a fine one, Museum curator W. D. Matthew offered one hundred dollars for it. The farmer was not pleased: "Wish to say in regard to the Matter, that I had expected to receive a better offer." In the end, he was smart enough to realize that no one else was likely to pay him as much, and his mammoth bones soon entered the Museum's collection.

THE LIVING ANIMAL

Like its close relatives, the modern elephants, *Mammuthus* almost certainly traveled in herds. Its teeth, designed for grinding coarse plant material, allowed it to feast on the tough grasses that thrived at the time.

Almost all mammoths died out about ten thousand years ago, although evidence reported in early 1993 shows that at least one species—a dwarf mammoth that inhabited Wrangel Island in the Arctic Ocean—seems to have survived until four thousand years ago. Scientists believe that the various species of *Mammuthus* were often hunted by early humans, and it may have been that over-hunting hastened the extinction of these great probiscideans in North America.

The Museum's spectacular *Megaloceras.*

MEGALOCERAS

Order: Artiodactylae
Family: Cervidae
Late Pleistocene
Discovered near Limerick, Ireland; acquired in 1872

While looking at fossil skeletons, it's easy to assume that all ancient animals lived—and became extinct—millions of years ago, during a time far removed from all human experience. As the magnificent extinct deer *Megaloceras* proves, however, by the standards of geologic time, some extraordinary animals survived almost to the present day.

Megaloceras thrived in Europe and Asia until about twelve thousand years ago, when their population mysteriously began to dwindle. Scientists believe the genus hung on for an additional ten thousand years, until it finally became extinct around 500 B.C. Ancient cave paintings in France and elsewhere depict animals that were probably *Megaloceras*. It is quite possible that early humans hunted this huge deer, perhaps even depending on its abundant flesh for food during the long Ice Age winters.

THE LIVING ANIMAL

At eight feet in length and seven hundred pounds in bulk, *Megaloceras* was a large animal by any standard. Its antlers were especially stupendous, spanning a breathtaking twelve feet and weighing as much as one hundred pounds. Like all deer, *Megaloceras* probably had to shed its antlers each year, growing a new rack in time for antler-wrestling duels that determined dominance over the family group or herd.

Moschops, and its bizarre and spectacular skull.

Broom's faith in the significance of the therapsids was borne out in his work; he was largely responsible for introducing the great South African beds of therapsid fossils to the world. His *Moschops* skeleton is just one of several finds in the American Museum's collection; his other discoveries grace the exhibit halls of museums worldwide.

THE LIVING ANIMAL

Moschops was an extraordinary creature, boasting a massive, barrel-shaped body, thick limbs tipped with tiny feet, and oddest of all, an enormously thick forehead and upper skull. Along with other skeletal features, this thickened cranium may have provided protection when two *Moschops* butted heads, perhaps in ritual battles for dominance over a family group or herd. Some paleontologists think that the "bonehead" dinosaurs, such as *Pachycephalosaurus*, may have engaged in similar behavior, as do bighorn sheep and other modern mammals.

MOSCHOPS

Subclass: Synapsida
Order: Therapsida
Suborder: Dinocephalia
Late Permian
Discovered in 1913 by Robert Broom in Spitzkop, South Africa

In 1932, the Scottish physician-paleontologist Robert Broom wrote, "The mammal-like reptiles of South Africa may be safely regarded as the most important fossil animals ever discovered." While others might disagree, no one can doubt that the therapsids, what Broom called mammal-like reptiles, include a host of fascinating animals that gave us great insights into the evolution of life on earth. Even now, the therapsids possess a resonance shared by few other ancient animals; as Broom recognized, "there is little or no doubt that among them we have the ancestors of the mammals, and the remote ancestors of man."

Paraceratherium

Paraceratherium

Order: Perissodactyla
Superfamily: Rhinoceratoidea
Family: Hyracodontidae
Oligocene
Discovered in 1922 by Wang, chauffeur to the Central Asiatic Expeditions in Urga, Mongolia

The unearthing of *Paraceratherium* in August 1922 was the first great triumph of Roy Chapman Andrews' expeditions to Mongolia. Yet, as so often is the case, the discovery of this magnificent early rhinoceros—known consecutively as *Baluchitherium* and *Indricotherium* before receiving its present-day name—was far more a matter of luck than design.

As Andrews wrote in his unpublished field journal, "Wang, one of the chauffeurs, had discovered it lying exposed in the bottom of a V-shaped gully in the bad land pocket which they had stopped to investigate on their way to camp. It seems to be always like that—the best things are found at the last moment."

Visiting the site, Andrews spotted exposed pieces of bone in the sand at the bottom of the gully. "I leaped down the slope & gloated over the great fragments protruding from the soil," he wrote. Andrews, Wang, and expedition photographer J. B. Shackelford immediately excavated portions of the skeleton, including a one-hundred-pound chunk of rock that appeared to contain the skull.

Throughout the excavation, Andrews was ecstatic, writing that "we tried to prospect further, but I was too excited from the find to stay. I wanted to get back to camp & spread the news." Returning to camp, "we found the men at tea and 'great was the rejoicing thereof.'"

The intoxication of the great discovery did not wear off quickly. "I could not go to sleep for hours—& in fact the whole camp quieted very slowly," he wrote in his journal. "And then *Baluchitheriums* were rambling thru' my dreams until morning."

THE LIVING ANIMAL

Fully living up to the excitement it provoked in Andrews, the spectacular *Paraceratherium* may have been the largest land mammal ever to live on earth. Some individuals reached more than twenty-five feet in length and weighed about fifteen tons, making them far larger than the most massive living elephant.

With its great size and long—although very heavy—neck, *Paraceratherium* could probably have browsed on treetop vegetation beyond the reach of any contemporary plant eater, an adaptation most closely mirrored in modern giraffes.

PLATYBELODON

Order: Proboscidea

Suborder: Elephantoidea

Family: Gomphotheriidae

Late Miocene

Several individuals discovered in 1930 by Albert Thomson, Walter Granger, and others in Wolf Camp, Inner Mongolia

"The mastodon quarry is certainly a remarkable one," Museum curator Walter Granger wrote in a letter to curator George Gaylord Simpson soon after the discovery of a mass grave of *Platybelodon*, bizarre, shovel-tusked elephants. "Apparently a sort of shovel-tusker convention met—and was overwhelmed."

Writing two years later in *The New Conquest of Central Asia*, Roy Chapman Andrews painted a more colorful picture of the ancient bog he called a *Platybelodon* deathtrap. "Out there in the desert in the brilliant sunlight in the year 1930, we were reading the story of a tragedy enacted millions of years ago," he recounted. Dramatizing the deaths, he described how the first victim "worked his way slowly along the shore....Suddenly, amidst his greedy feeding he found he could not lift his ponderous leg. He struggled madly only to sink deeper

Platybelodon

and deeper into the mire of death....Another came and still others, each one to die as he had died."

THE LIVING ANIMAL

Platybelodon's remarkable feeding apparatus featured shortened upper jaws with two short, downward-pointing tusks. The bottom jaw, by contrast, was extraordinarily long and broad and contained the flat, wide "shovel tusks."

Scientists believe that this arrangement of tusks represented an adaptation for specialized feeding. Most likely, *Platybelodon* lived in and around lakes, streams, or other wetlands. Bending down, it dug up soft plants with its lower tusks, and gripping the vegetation between the lower jaws and flattened tusk, the animal then wrenched its food out of the ground by raising its head.

STENOMYLUS

Order: Artiodactyla
Family: Camelidae
Early Miocene
Discovered in 1907, near Agate Springs, Nebraska; collected in 1908

Writing to Henry Fairfield Osborn in 1908, W. D. Matthew said that the *Stenomylus* quarry near Agate Springs "appears to be a very extensive one likely to yield hundreds of articulated skeletons, if worked deep enough." So it was no problem for the American Museum's collectors to unearth the nine skeletons of this little Miocene camel that now grace the mammal halls.

Instead of simply leaving the skeletons in situ or mounting them all in lifelike postures, the Museum's preparators, under the guidance of Barnum Brown and Walter Granger, chose a more creative course. While in Patagonia, Brown had seen guanacos (relatives of camels) bedding down during severe winter weather. At his suggestion, the *Stenomylus* exhibit was

Stenomylus

mounted to show four animals lying dead from starvation or cold, and five individuals in various lifelike poses, such as lying down and rising to their feet. Looking at this clever display, it's easy to imagine what life would have been like in Miocene North America, when the cold winter wind whipped across the open plains.

THE LIVING ANIMAL

Small—no more than about three feet in length—and delicately built, *Stenomylus* was shaped more like a modern gazelle than the heavy-bodied, awkward modern camels. It may have lived like a gazelle too, gathering in small groups or herds and relying on its speed and agility to avoid the wolves and other carnivores that would have preyed on it. *Stenomylus'* well-developed incisors (its lower canines and first premolar teeth were shaped like incisors as well) were ideal for cropping the low vegetation that covered the Great Plains of North America.

TOXODON

Order: Notoungulata
Suborder: Toxodonta
Family: Toxodontidae
Pleistocene
Discovered near Buenos Aires, Argentina; acquired in 1910

Looking upon the bones of *Toxodon* during his famous 1831 to 1836 voyage to South America, Charles Darwin was moved to amazement. "Toxodon, perhaps one of the strangest animals ever discovered," he wrote. "In size it equalled an elephant or megatherium; but the structure of its teeth…proves indisputably that it was intimately related to the Gnawers, the order which at the present day includes most of the smallest quadrupeds. In many details it is allied to the Pachydermata. Judging from the position of its eyes, ears, and nostrils, it was probably aquatic, like the dugong and manatee, to which it is also allied. How wonderfully are the different orders, at the present time so well separated, blended together in different points of the structure of the toxodon!"

This large, heavy-bodied foliage feeder was just one of many strange pampean mammals of South America. Here too were the bizarre *Macrauchenia*, the ground sloth *Glossotherium*, the great sabertooth cat *Smilodon*, and many others, which together formed one of the most fascinating faunas ever to evolve on earth.

Toxodon head, and restored skull and jaws.

THE LIVING ANIMAL

Boasting a heavy body and a broad, blunt head, nine-foot-long *Toxodon* most closely resembled a modern rhinoceros minus the horn. Its complex dental structure, comprising chisel-like incisors and high-crowned, curved cheek teeth that grew throughout the animal's life, suggests that *Toxodon* was well adapted to feast on coarse Pampas grasses.

GALLERY OF OTHER ANCIENT CREATURES

AXELRODICHTHYS

Class: Osteichthyes
Subclass: Sarcopterygii
Order: Actinista
Family: Coelancanthidae
Cretaceous
Discovered in Araripe Plateau, Brazil

The Araripe Plateau, an extraordinary expanse of land set in a little-visited region of northeastern Brazil, is home to the Santana Formation, one of the richest Cretaceous fossil sites on earth. For nearly two centuries, explorers, scientists, and local farmers have unearthed a steady stream of fossil fish, crocodilians, turtles, pterosaurs, and other fossil animals, many spectacularly well preserved. In recent years, Museum curator John G. Maisey and others have catalogued the dozens of genera found so far, producing an indelible portrait of life one hundred and ten million years ago in what was then a rich coastal and marine environment. Several of the Santana Formation fish, including *Axelrodichthys*, are on display in the Museum's hall of primitive vertebrates.

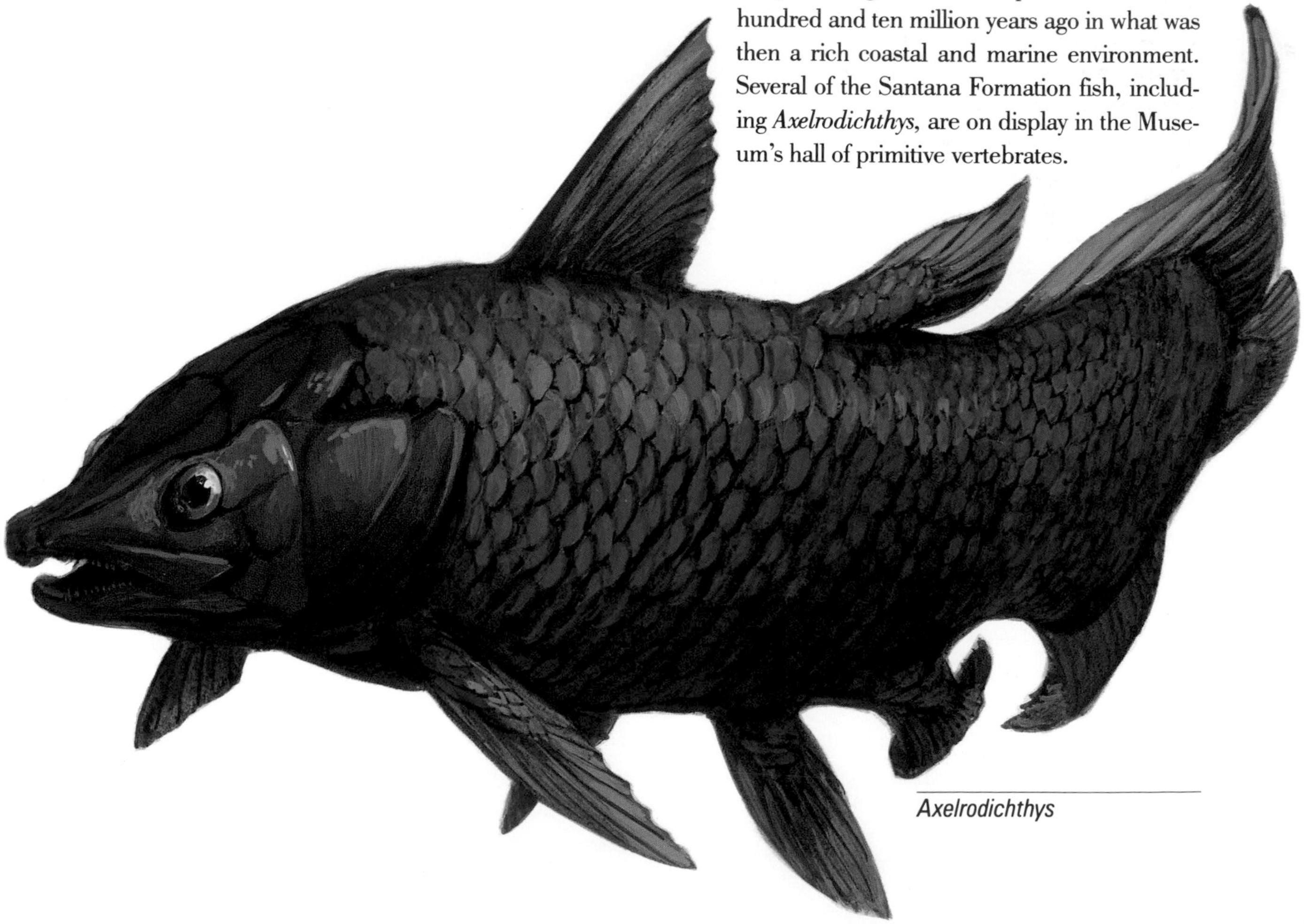

Axelrodichthys

Axelrodichthys belongs to one of the strangest and most remarkable families of fish yet known: the coelacanths. Until the 1930s, scientists believed that the last members of this once-widespread family had disappeared at the end of the Cretaceous period, about sixty-five million years ago. But in 1938, fishing boats captured living coelacanths—proving beyond a doubt that the family had somehow survived, without leaving any fossil evidence, for tens of millions of years longer than previously believed.

THE LIVING ANIMAL

At the height of their abundance and diversity, coelacanths occupied many different ecological niches, from open-ocean environments to shallow-water swamps. Heavy-bodied *Axelrodichthys*, which may have reached six feet in length, lived in clear, coastal waters. Perhaps it fed on smaller fish that feasted on the leaves, coniferous cones, and other plant material falling into the water from the nearby forested shoreline.

COLOSSOCHELYS

Order: Chelonia
Suborder: Cryptodia
Family: Testudinidae
Pleistocene
Discovered in 1922 by Barnum Brown in Siwalik Hills, India

Barnum Brown, the Museum's expert fossil collector, discovered the magnificent fossil remains of *Colossochelys*, the largest tortoise yet known, at the onset of his ambitious 1922–1923 collecting expedition to India. At first, however, Brown wasn't certain that the specimen was even worth collecting: It had weathered out of the rock and lay scattered in thousands of pieces, none larger than a human hand. The fragmented skeleton, Brown wrote, was "mute testimony that 'dust thou art and unto dust thou shalt return.'"

Colossochelys, magnificently restored from its jigsaw puzzle origins.

Filled with a fossil hunter's certainty that finer specimens were just around the bend, Brown left this find uncollected. After a year and a half of prospecting, however, he had failed to discover another *Colossochelys* specimen that was nearly as complete, leaving him no alternative but to undertake the tedious job of collecting the fractured skeleton he had left behind.

Back at the Museum, preparator Otto Falkenbach worked on the specimen for a year. "With the thousands of pieces it was literally a picture puzzle," Brown recalled. And even when the restoration was completed, the shell proved to be badly distorted. Preparators had to break it back into pieces and reconstruct it, finally completing the free-mounted skeleton in

1931—a full eight years after the tortoise first arrived at the Museum.

THE LIVING ANIMAL

The largest *Colossochelys* may have reached eight feet or longer and weighed more than one ton. Scientists believe that the giant tortoise probably ate vegetation, cropping its food with its toothless beak, much as its modern counterparts do. Present-day giant tortoises, such as those that survive on the Galapagos Islands, have been known to live longer than one hundred years, leading experts to suppose that *Colossochelys* could also have lived to such a great age.

Cryptoclidus

CRYPTOCLIDUS

Order: Plesiosauria
Superfamily: Plesiosauroidea
Late Jurassic
Discovered in Peterborough, England; acquired in 1902

Cryptoclidus was a mid-size plesiosaur, one of a group of marine reptiles that thrived in the world's oceans from the Early Jurassic through Late Cretaceous times—an astonishing total of more than one hundred million years. Although descended from land-based reptiles, *Cryptoclidus* and other plesiosaurs were superbly adapted to marine life. Propelling themselves with figure-eight or crescent-shaped motions of their long, flexible paddles, plesiosaurs could "fly" through the water, much as penguins do today.

Most scientists believe that plesiosaurs lacked one adaptation to life underwater. Like modern sea turtles, they were unable to give birth at sea and had to drag themselves onto a beach to deposit their eggs.

THE LIVING ANIMAL

Like other predators of its time, long-necked *Cryptoclidus*, which reached about thirteen feet in length, seemed almost perfectly adapted for hunting. Its extraordinarily strong, flexible front paddles would have enabled it to move quickly and powerfully through the Jurassic seas; its long neck gave it agility; and its deep, muscular skull and jaws would have allowed it to snap up the fish and other marine creatures that were its prey.

Cryptoclidus possessed abundant sharp, curved teeth that interlocked when its jaws were closed. This feature has led some scientists to believe that *Cryptoclidus* may have snatched at schools of small fish or other marine life, trapping its prey behind a mesh of closed teeth before swallowing them.

Cryptoclidus skeleton.

DIADECTES

Order: Diadectomorpha
Family: Diadectidae
Early Permian
Discovered in 1906 by E. C. Case in Archer County, Texas

Early in this century, E. C. Case, one of the most talented collectors of primitive reptiles and amphibians and a fossil-hunting professor at the University of Michigan, was loaned to the American Museum for a single field season.

Diadectes

Working in the Permian fossil beds of Texas, he sent back specimens of the enormous salamander *Eryops*, the remarkable synapsid *Dimetrodon*, with its famous sail of skin and bone, and *Diadectes*, which was perhaps the most unusual of all.

Diadectes, with its bulky body, thick legs, and strong jaws filled with grinding teeth, appears to have stood on the evolutionary border between reptiles and amphibians. Today, most experts acknowledge its skeletal similarities to modern reptiles, but they still believe that, overall, *Diadectes* was also closely related to the amphibians.

THE LIVING ANIMAL

Ten-foot-long *Diadectes* was one of the largest land animals to inhabit North America during the Early Permian period. It also appears to have been one of the first land-based animals to eat plants—evidence of a preliminary attempt at a way of life that would eventually allow terrestrial herbivores to spread explosively across the continents.

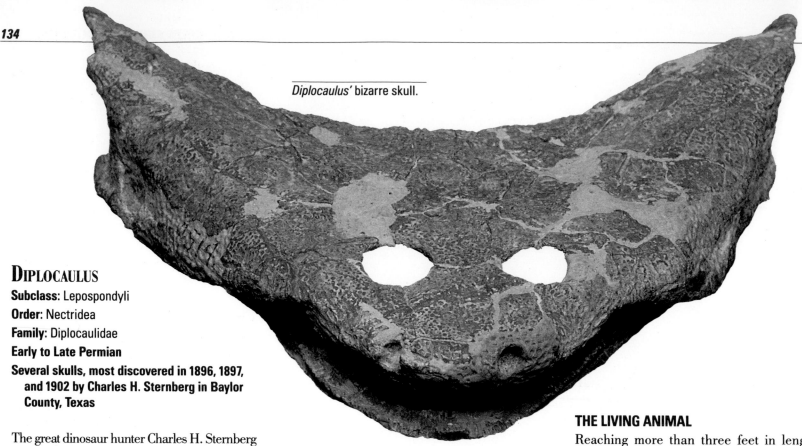

Diplocaulus' bizarre skull.

DIPLOCAULUS

Subclass: Lepospondyli
Order: Nectridea
Family: Diplocaulidae
Early to Late Permian
Several skulls, most discovered in 1896, 1897, and 1902 by Charles H. Sternberg in Baylor County, Texas

The great dinosaur hunter Charles H. Sternberg was well known for the unflagging enthusiasm he brought to his discoveries. Whether he was gazing upon the great "mummy" *Edmontosaurus* or a tiny, seemingly insignificant ancient animal, Sternberg always thanked God, fate, and Henry Fairfield Osborn, the Museum's curator of vertebrate paleontology, for allowing him to pursue his lifelong passion.

Diplocaulus, an odd-looking early amphibian, was far from Sternberg's rarest or most spectacular find. It was a common denizen of its proper habitat (the Permian Red Beds of Texas), and today the American Museum displays several of Sternberg's *Diplocaulus* skulls, including an extensive series that charts the amphibian's growth.

The animal's relative abundance and small size meant that Sternberg, always hard-pressed for funds, couldn't expect to earn much for even a perfect specimen. Worse, the best fossils were often found in the most remote, parched sections of the Texas badlands. "How can I describe the hot winds, carrying on their wings

clouds of dust?" Sternberg recalled in his 1909 memoir, *The Life of a Fossil Hunter.*

But to Sternberg, such obstacles were made to be overcome. After extensive hunting, he came upon a preserved ancient pond bed ("the very place to look for fossils") and soon discovered it was replete with *Diplocaulus* and other Permian creatures. Ever enthusiastic, Sternberg recalled in his memoirs his emotions: "'This,' my notes say, 'promises to be one of the finest localities I have found, and pays for the days of search under trying conditions.'"

Diplocaulus' vertebrae.

THE LIVING ANIMAL

Reaching more than three feet in length, *Diplocaulus* was an inhabitant of lakes and streams. While its body was reminiscent of a modern salamander's, its head featured a pair of "horns" extending several inches from each side of the head in adults.

Skeletal structures such as *Diplocaulus'* horns frustrate paleontologists, who have a natural impulse to propose possible uses for such features. Some guess that the horns were used to make *Diplocaulus* seem larger than it was, thereby discouraging predators from attempting to swallow its head.

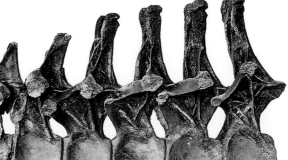

DUNKLEOSTEUS

Class: Placodermi
Order: Arthrodira
Family: Dinichthyidae
Late Devonian
Skeleton acquired from Cleveland

This extraordinary fish, part of a group known as the arthrodires, was one of the largest of its time. Reaching more than ten feet in length, it featured heavily armored bony plates that interlocked to create an impervious head shield, further plates on its upper back and sides, and a tapering, scaleless body edged with fins. Most likely it wriggled its body in eel-like fashion to chase down its prey, which may have included large fish, squids, and other marine creatures.

Cosmopolitan in its distribution, *Dunkleosteus* lived at the same time and in the same oceans as such ancient sharks as *Cladoselache* (also on display in the halls). However, this enormous fish would have had little to fear from the six-foot *Cladoselache*—or any other inhabitant of the Late Devonian seas.

THE LIVING ANIMAL

An unusual adaptation of *Dunkleosteus* and many other placoderms was their lack of teeth. Instead, *Dunkleosteus* featured huge dental plates equipped with toothlike spikes at the front and sharp cutting surfaces farther back. Powered by strong muscles, these dental plates would have allowed *Dunkleosteus* to grind up and ingest even the best-protected prey.

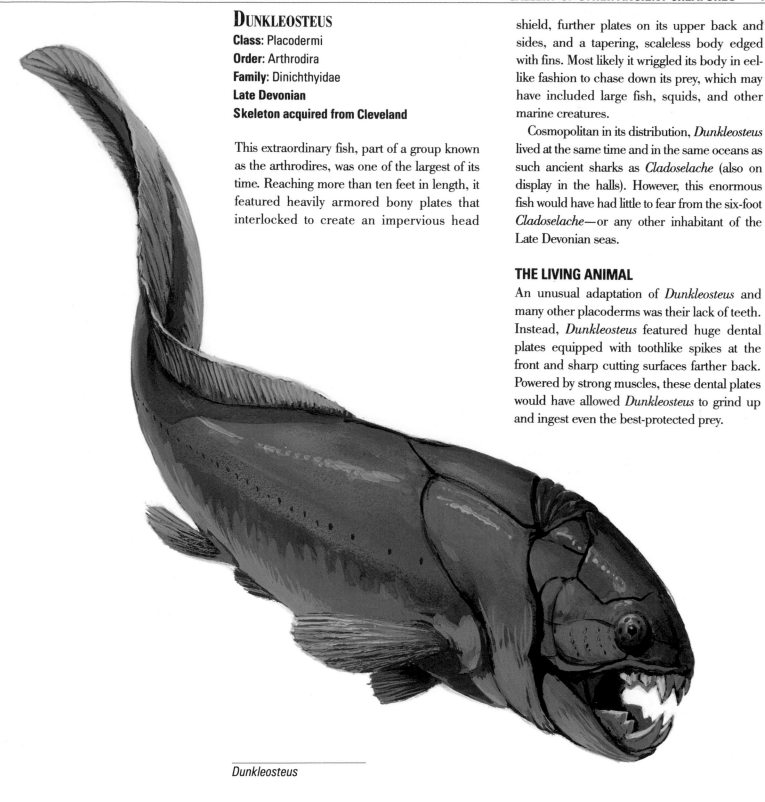

Dunkleosteus

PROGANOCHELYS

Order: Chelonia
Suborder: Proganochelydia
Family: Proganochelyidae
Late Triassic
Discovered in Trössingen, Germany; cast acquired in 1929

The Museum's *Proganochelys,* the oldest well-preserved turtle yet discovered, was unearthed from the same Trössingen quarry that produced the Museum's more famous specimen of *Plateosaurus,* the Late Triassic sauropod. From the proximity of their remains, it is believed that the primitive turtle and the dinosaur shared the same haunts (apparently a lake and its edges), along with early sharks and other creatures.

Displayed in the Stuttgart Museum before World War II, the *Proganochelys* skeletons almost didn't survive the war. As American Museum paleontologist Eugene Gaffney described the skeletons' history in a 1990 Museum *Bulletin* article, "The German government proclaimed any evacuation of collections to bombproof shelters as defeatist and imposed severe penalties but [Chief Curator] Berckhemer was able to have most of the collections safely removed to other locations. This was fortunate because the Stuttgart Museum building was completely destroyed by bombing."

Today, the important original *Proganochelys* specimens reside in the Stuttgart Museum's rebuilt quarters. The individual on display at the American Museum is a cast.

THE LIVING ANIMAL

More than three feet in length, *Proganochelys* bore a strong resemblance to the modern snapping turtle. Scientists studying *Proganochelys'* skeleton, however, are able to note many differences from today's turtles, including several features that tie *Proganochelys* to its earlier

Trössingen landscape in the Triassic: the turtle *Proganochelys* with hyobodont shark and the dinosaur *Plateosaurus.*

relatives, which were most likely shell-less anapsid reptiles called captorhinids.

Proganochelys probably lived in lakes and may have fed on vegetation or small fish. It would have had to climb onto land in order to breed, as do all modern turtles.

PROTOSUCHUS

Order: Crocodylia
Suborder: Protosuchia
Family: Protosuchidae
Early Jurassic
Discovered in 1930 in Cameron, Arizona, and collected in 1931 by Barnum Brown

The first fossil remains of this early relative of the crocodile were discovered in 1930 by a Navajo man living in the Arizona desert. News of the find, however, did not leak out of this sparsely populated region for quite some time—until the tidbit of information reached the local trading post. To the Museum's good fortune, the proprietor was a friend of the fossil hunter Barnum Brown.

After one look at the fossils, the shrewd man contacted Brown, who journeyed to the site the following summer. In less than two weeks, Brown had collected four skeletons of *Protosuchus*, including the magnificent specimen on display in the Museum's halls.

THE LIVING ANIMAL

Crocodilians are considered among the most successful animals ever to exist on earth. The first members of the order appeared late in the Triassic period, at the same time as the earliest dinosaurs. Yet, while all the dinosaurs except for birds have been extinct for sixty-five million years, alligators and crocodiles still thrive in appropriate habitat. Even more remarkably, today's crocodilians look quite similar to some of their ancestors that lived more than two hundred million years ago.

Protosuchus, which reached about three feet in length, shared many characteristics, such as armored scales, with modern crocodiles. But while modern crocodilians always live in or near water, *Protosuchus* lived on dry land, sharing the southwestern landscape with coelurosaurs, carnosaurs, and other Jurassic dinosaurs.

Protosuchus life restoration, and one of the specimens found by Barnum Brown.

PTERANODON

Order: Pterosauria
Suborder: Pterodactyloidea
Family: Pteranodontidae
Late Cretaceous
Three specimens, discovered in 1894 and 1916 by H. T. Martin and in 1952 by Bobb Schaeffer and Carl Sorenson, all in Kansas

Othniel Charles Marsh, a dominant personality in nineteenth-century paleontology, was the first to discover fossils of pterosaurs in North America. He was also the first to locate the mecca of pterosaur fossils that lay buried in the chalk beds of western Kansas.

In the decades since Marsh's initial find in 1870, paleontologists have unearthed remains of at least a half dozen species of *Pteranodon* in the Late Cretaceous Kansas chalk beds, and other North American pterosaur fossils have turned up in Texas, Montana, Alberta, and elsewhere. While *Pteranodon*, with its twenty-three-foot wingspan, was among the largest of the flying reptiles, it paled in comparison to another North American pterosaur: the great *Quetzalcoatlus*, whose wings spanned a breathtaking forty feet.

THE LIVING ANIMAL

Despite its great wingspan (far larger than any modern bird's), *Pteranodon* had extremely lightweight, hollow bones, which allowed the huge reptile to weigh less than forty pounds. Over the years, paleontologists have debated *Pteranodon's* flying ability. Some argue that this pterosaur was a weak flyer; it may have used its long, skin-covered wings only to glide in a convenient breeze, as modern albatrosses do. An increasing number of scientists now believe, however, that a more accurate modern comparison to *Pteranodon* is the gull, which is capable both of soaring and of sustained, active flight.

Pteranodon had a hatchet-shaped skull and toothless jaw. Most experts believe that these features equipped *Pteranodon* for catching and eating fish; perhaps it hunted by scooping up its prey and swallowing it whole, as do today's pelicans.

RHAMPHORHYNCHUS

Order: Pterosauria
Suborder: Rhamphorhynchoidea
Family: Rhamphorhynchidae
Late Jurassic
Found in Solnhofen, Germany; acquired in 1922

Apparently a common pterosaur but confined to a limited range, *Rhamphorhynchus* roamed the skies above Europe about one hundred and fifty million years ago. The best-known specimens come from the astounding Solnhofen limestones in southern Germany—the same site that produced the first pristine *Archaeopteryx* skeleton, complete with feather impressions. Similarly, Solnhofen *Rhamphorhynchus* skeletons sometimes show delicate impressions of the skin membranes that stretched along the bones of the wings.

The *Rhamphorhynchus* individuals preserved in the Solnhofen limestones were among the last of the *Rhamphorhynchidae*, a family that had survived since the Late Triassic period, seventy million years earlier. By the end of the Jurassic period, the dominant pterosaurs were the larger, more advanced pterodactyls, which survived until the end of the Cretaceous—when all pterosaurs became extinct.

THE LIVING ANIMAL

Typical of its family, *Rhamphorhynchus* was small, only about seven inches long with a wingspan reaching about three feet. It probably held its long, slender tail, which was tipped with a diamond-shaped vane, stiffly behind it to counterbalance its comparatively heavy neck and head.

Rhamphorhynchus' mouth held outward-pointing teeth that meshed when the narrow jaws closed. Scientists believe that this pterosaur ate fish; in fact, one of the marvelous Solnhofen specimens died with an undigested fish still in its stomach.

Pteranodon

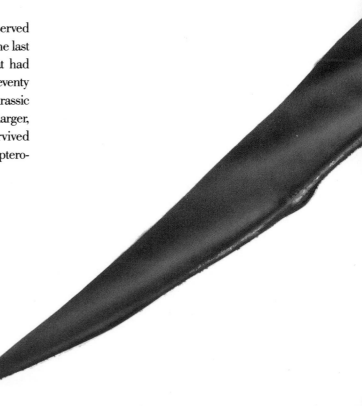

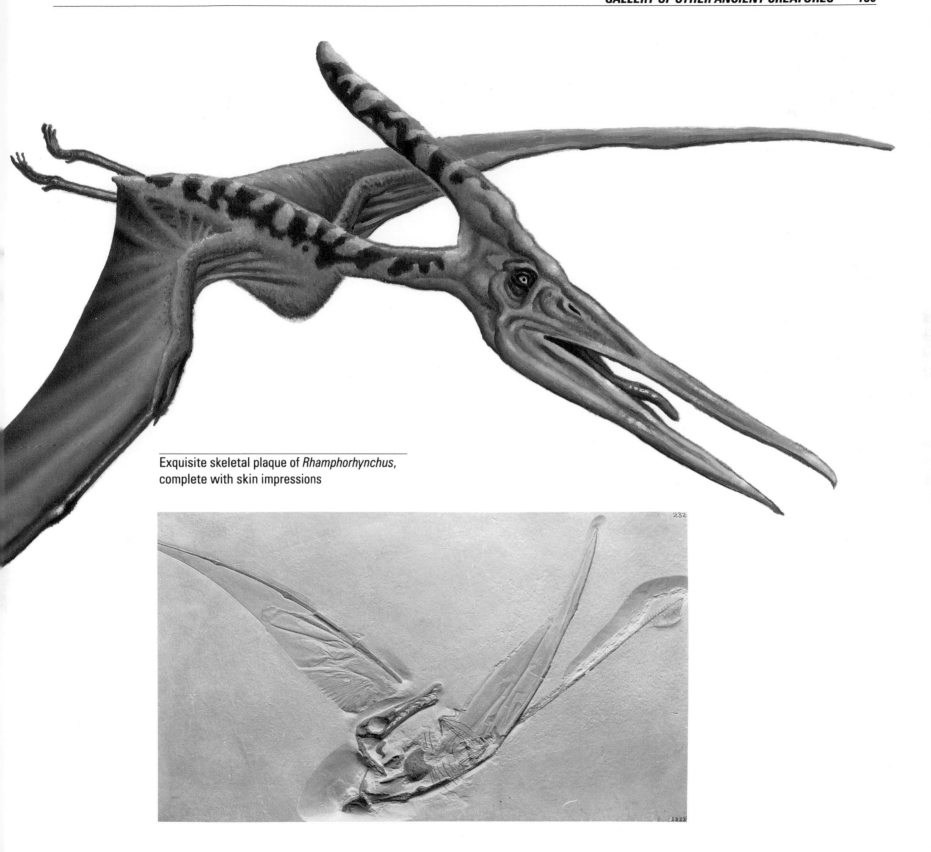

Exquisite skeletal plaque of *Rhamphorhynchus*,
complete with skin impressions

RUTIODON

Superorder: Archosauria
Suborder: Phytosauria
Late Triassic
Discovered in 1895 by W. D. Matthew in Egypt, North Carolina

Not surprisingly, paleontologists' ears prick up when they hear of an opportunity to go underground with a minimum of digging. Tell a fossil hunter you've discovered an unexplored cave, and chances are the cave won't remain unsearched for long.

Even more fascinating to paleontologists than caves are mines, which can provide entry to otherwise inaccessible fossil beds hundreds of yards below ground. In June 1895, W. D. Matthew was lured to Egypt, North Carolina, by the promise of Triassic fossils in the local coal mine.

Trying to excavate bones from a working mine, however, can be a tricky, even risky, business, but Matthew's letters to Henry Fairfield Osborn, curator of vertebrate paleontology at the Museum, show that he understood the risk but was not cowed by it. "There has been a general shake-up in the mine," he wrote matter-of-factly. "Mr. Williams [a supervisor] was killed in an explosion, shortly after I left last summer."

The miners could be of great help, although sometimes their tales grew taller than reality. "Tyson [a miner] tells me of bones having been found last winter," Matthew reported. "He spoke of them as if there were a great many of them—which I take with sundry grains of salt." Such reports, however inflated, could still provide a legitimate lead. "It really appears, though, from what one of the other men said, that he found one bone there, at least."

In this particular case, the lead proved fertile. Within a week of searching at the miner's site, Matthew had excavated a prize find. "Of the saurian remains I have made some very good hauls," he wrote Osborn. "The best so far is what appears to be a head, with jaws over two feet long."

Matthew soon found that the skull and jaws were associated with a slew of other bones. When reconstructed, they formed a beautiful skeleton of the primitive carnivore *Rutiodon*.

As well as being an important fossil, the Museum's *Rutiodon* carries another distinction. It was the very first fossil in the Museum's catalog of ancient reptiles: number one in a collection that would soon include *Tyrannosaurus*, *Apatosaurus*, *Protoceratops*, *Pteranodon*, and many other well-known creatures.

THE LIVING ANIMAL

Rutiodon, like other phytosaurs, superficially resembled a crocodile. The groups, however, were not closely related, even though they evolved similar characteristics, including body armor and long, toothy jaws, to suit their similarly swampy environments.

At ten feet in length, *Rutiodon* was one of the most powerful predators inhabiting the North American and European wetlands during its time. Its diet may have included fish, other reptiles (the remains of which have been found in the stomachs of fossilized *Rutiodons*), and even the small dinosaurs that lived alongside it in the Late Triassic period.

Rutiodon

Stenopterygius

STENOPTERYGIUS

Order: Ichthyosauria
Family: Stenopterygiidae
Early to Mid-Jurassic
Several individuals discovered in Württemberg, Germany; acquired in 1896, 1901, and 1907

The ichthyosaurs, carnivorous reptiles that inhabited the world's oceans, evolved into forms quite similar to modern dolphins. All had streamlined bodies, paired paddles, strong, crescent moon–shaped tail fins, and slender beaks lined with sharp teeth.

An extraordinary trove of Early Jurassic fossils unearthed from an ancient seabed in southern Germany included perfectly preserved specimens of *Stenopterygius* and other ichthyosaurs. Many (including one of the Museum's skeletons) have unborn young still inside their bodies, proving that unlike other marine reptiles, ichthyosaurs bore live young at sea. Remarkably, a few individuals found at this site actually died while giving birth; the young were in the midst of emerging, tail first, at the precise moment of the mother's death.

THE LIVING ANIMAL

Like other ichthyosaurs, *Stenopterygius* (which reached about ten feet in length) swam by beating its tail from side to side; its paddles were employed to change direction. Its streamlined shape and strongly muscled tail most likely allowed *Stenopterygius* to chase down and eat fish and other marine animals.

SELECTED BIBLIOGRAPHY

Archival Material

The author has relied on diaries, field journals, correspondence, and other unpublished writings by Roy Chapman Andrews, Roland T. Bird, Barnum Brown, Edwin H. Colbert, Walter Granger, W. D. Matthew, Rachel Husband Nichols, Henry Fairfield Osborn, George Gaylord Simpson, Charles H. Sternberg, and many others. All of this material can be found in the Rare Book Room and the Archives of the Department of Vertebrate Paleontology, American Museum of Natural History, New York City.

Journals and Magazines

American Museum Novitates. New York: 1923—.

Bulletin of the American Museum of Natural History. New York: 1881—.

Galusha, Theodore. "Childs Frick and the Frick Collection of Fossil Mammals." *Curator*, January 1975.

Natural History magazine. New York: 1919—.

Books

Andrews, Roy Chapman. *This Business of Exploring.* New York: Putnam, 1935.

———. *Under a Lucky Star.* New York: Viking, 1943.

———, et al. *The New Conquest of Central Asia.* New York: American Museum of Natural History, 1932.

Benton, Michael. *Dinosaur and Other Prehistoric Animal Fact Finder.* New York: Kingfisher, 1992.

Bird, Roland T. *Bones for Barnum Brown.* Fort Worth: Texas Christian University, 1985.

Brown, Lillian. *I Married a Dinosaur.* New York: Dodd, Mead, 1950.

Colbert, Edwin H. *Digging into the Past.* New York: Dembner, 1989.

———. *A Fossil-Hunter's Notebook.* New York: Dutton, 1982.

———. *The Great Dinosaur Hunters and their Discoveries.* 1968. Reprint. New York: Dover, 1984.

Dixon, Dougal, et al. *The Macmillan Illustrated Encyclopedia of Dinosaurs and Prehistoric Animals.* New York: Macmillan, 1988.

Gaffney, Eugene S. *Dinosaurs.* New York: Golden, 1990.

Gross, Renie. *Dinosaur Country.* Saskatoon, Saskatchewan: Western Producer Prairie, 1985.

Lambert, David. *A Field Guide to the Dinosaurs.* New York: Avon, 1983.

Maisey, John G., et al. *Santana Fossils: An Illustrated Atlas.* Neptune, N.J.: T.F.H., 1991.

Norell, Mark. *All You Need to Know About Dinosaurs.* New York: Sterling, 1991.

Osborn, Henry Fairfield. *The AMNH.* New York: The Irving Press, 1911.

Perkins, John. *To the Ends of the Earth.* New York: Pantheon, 1981.

Preston, Douglas J. *Dinosaurs in the Attic.* New York: St. Martin's, 1986.

Rainger, Ronald. *An Agenda for Antiquity.* Tuscaloosa, Ala.: The University of Alabama, 1991.

Sattler, Helen Roney. *The New Illustrated Dinosaur Dictionary.* New York: Lothrop, Lee & Shepard, 1990.

Simpson, George Gaylord. *Concession to the Improbable.* New Haven, Conn.: Yale University, 1978.

Sternberg, Charles H. *Dinosaur Hunting on Red Deer River, Alberta, Canada.* Lawrence, Kans.: 1917.

———. *Life of a Fossil Hunter.* New York: Henry Holt, 1909.

Wilford, John Noble. *The Riddle of the Dinosaur.* New York: Alfred A. Knopf, 1986.

INDEX

Agate Springs (Nebraska), 44–45, 127

Alaska, 43, 74

Alberta (Canada), 32–33, 36, 38, 44, 93, 95, 109, 139

Albertosaurus, 33, 40, 92–93, *93*

Alexander, John, 11, 75, *76*, 77

Alligators, 28, 31, 137

Allosaurus, 26, *26*, *27*, 74, 86–87, 90–91

Altangerel, Perle, 79, 80, 82, 83

American-Mongolian Expedition, 77–83, *78–79*

American Museum of Natural History, *20*

  Department of Vertebrate Paleontology, 17, 19, 21, 23, 41, 42, 50, 64, 65, 67, 72, 74, 75, 88, 101

  Frick Building, 43, 74, 75

  Hall of Human Biology and Evolution, 77

  Hall of Primitive Vertebrates, 87, 129

  opening, 17–18

  renovation of, 74, 87

  Roosevelt Rotunda, *86–87*

  storage areas, 74–75, *74–75*

  vertebrate paleontology halls, 9

Amphibians, 18, 133, 134

Anatosaurus, 38, 100

Anatotitan, 10, 94–95, *94*

Andrews, Roy Chapman, 10, 11, 43, 47–61, *48*, *51*, *52*, *54–55*, *59*, 63, 64, 72, 74, 79, 81, 82, 83, 103, 106, 112, 125

Andrewsarchus, 57, *57*, 74, 112–113, *113*

Ankylosaurus, 24, *24*, 33, 92, 95, 95

Antarctica, *70–71*, 71

Apatosaurus, 10, 18, 25, *25*, 26, *27*, 66, 88, 96, *97*, 105, 107, 140

Archaeopteryx, 82, 83, 88, 98, *98*, 105, 139

Axelrodichthys, 129–130, *129*

Baluchitherium, 52, *52*, 77, 125

Barbour, E. H., 44

Barosaurus, 66, 86–87, 90–91

Baryonyx, 105

Berkey, Charles P., 50, 52

Berman, David, 96
Bird, Roland T., 64–67
Birds, 10, *10*, 11, 18, 28, 31, 52, 82, 88, 98, 115
Bison, 74, 109
Bone Cabin Quarry (Wyoming), 25–27, *25*, *26*, 72, 107
"Bone Wars," 18
Bridger Basin (Wyoming), 77
Brontops, *90–91*, 113-114, *113*, *114*
Brontosaurus, 18, 25, 96, 107
Broom, Robert, 124
Brown, Barnum, 10, 11, 19, *21*, 23–35, *24*, *28*, 36, 40, 44, 47, 64, 67, 68, 72, 74, 80, 92, 94, 95, 104, 107, 108, 109, 127, 137

California, 44, 117
Camarasaurus, 96, 107
Canis dirus, 115, *115*
Carnivores, 9, 10, 26, 28, 66, 67, 74, 93, 105, 112, 113, 115, 117, 141
Carnosaurs, *11*, 109, 117
Case, E. C., 116, 117, 133
Cenozoic period, 43
Central Asiatic Expedition, 47–61, 63, 74, 81, 125
Ceratopsians, *58*, 59, 103, 104, 105, 109
Chiappe, Luis, 82, 83
Chinle Formation, 68–69
Christman, Charles, 19
Cladistics, 88
Cladoselache, 135
Clark, James, 79, 81, 82, 83
"A Classification of the Mammals" (Simpson), 114
Coelophysis, 68, *68*, 69, *69*, 71, 98, 99, *99*
Colbert, Edwin H., 50, 67, 68, *70–71*, 71, 99, 116
Colossochelys, 130, *130*
Como Bluff (Wyoming), *21*, 25–27, 32, 44, 96, 107
Compsognathus, 105
Concession to the Improbable (Simpson), 21, 50, 65
Continental drift, 71

Cope, Edward Drinker, 18, 19, 38, 94, 99, 109
Corythosaurus, *32*, 33, 92, 93
Cotylorhynchus, 117
Creodonts, 112
Cretaceous period, 19, 24, 28, 29, 32, 52, 57, 66, 69, 81, 83, 88, 92, 93, 94, 95, 100, 103, 104, 106, 108, 129, 131, 139
Crooked Creek (Montana), 94
Cryptoclidus, 131, *131*, *132*
Cuba, 24, 28, 31

Dakota Formation, 69
Darwin, Charles, 41, 98, 128
Deinodon, 92
Deinonychus, 98, 108
Devonian period, 135
Diadectes, 133, *133*
Diceratherium, 44–45, 45
Digging into the Past (Colbert), 67
Dimetrodon, 10, 116–117, *116*, 133
Dingus, Lowell, 79, 81, 87, 88
Dinohippus, 41
Dinohyus, *44–45*
Diplocaulus, 134, *134*
Diplodocus, 18, 26, 96, 107
Dogs, 115, 117
Dolphins, 117, 119
Dromaeosaurus, 82
Dunkleosteus, 135, *135*

Echolocation, 119
Edaphosaurus, 117
Edmontosaurus, 38, 40, 93, 95, 100–101, *100–101*, 109
Eggs, 52, *58*, 59, *59*, 74, 103, 106, *106*
Elephants, 9, 122, 126
Elliot, David, *70*
Enaliarctos, 117, *117*
Entelodon, 112
Entrada Formation, 69
Eocene period, 57, 76, 77, 112, 118
Eohippus, 41
Equus, 40, 41, 42, 118, *118*
Erlikosaurus, 81
Eryops, 133
Estesia, 81, *81*
Eurhinodelphis, 119, *119*

Evolution, 18, 21, 40, 41, 42, 64, 68, 77, 88, 113, 114
missing link, 49, 60, 98
Excavation techniques, 80

Falkenbach, Charles, 19
Falkenbach, Otto, 19, *24*, 130
Fish, 129
Flaming Cliffs of Shabarakh Usu. *See* Shabarakh Usu.
A Fossil-Hunter's Notebook (Colbert), 99
Fraley, Phil, 89, *89*
Frick, Childs, 43, *43*, 74

Gaffney, Eugene, 77, 136
Germany, 98, 105, 136, 141
Ghost Ranch (New Mexico), 62–63, 68, 69, 99
Gidley, James W., 19, 42, 118
Glen Rose trackways, 64–67, *64*, *65*, *66*
Glossotherium, 128
Gobi Desert (Mongolia), 10, 11, 47–61, 77–83
Granger, Walter, 19, *25*, 26, 36, 42, 50, 52, *52*, 56, 59, 60, 65, 66, 68, 71, 96, 103, 104, 106, 107, 112, 113, 126, 127
The Great Dinosaur Hunters and Their Discoveries (Colbert), 50

Hadrosaurs, 32–33, 34, 57, 93, 94–95, 100–101, *100–101*
Hell Creek (Montana), 28, 31, 32, 44, 72
Herbivores, 95, 103, 107, 108, 113, 117, 119, 125–126, 133
Hermann, Adam, 19
Holland, W. J., 96, 101
Hominids, 49, 50
Homo sapiens, 48
Homotherium, *115*
Hoplitosaurus, *24*
Horses, 40, 41–42, *41*, 43, 44, *44–45*, 74, *88*, 115, 118
Hylaeosaurus, 105
Hyopsodus, 77
Hypohippus, 42
Hyracotherium, 40, 41, 42, 118

Iguanodon, 105, 108, 117
Indricotherium, 125
Ippolito, Frank, 77
Iren Dabasu (Mongolia), 50, 52, 57, 60

Jesup, Morris K., 18, *18*
Johnson, Albert, 56
Jurassic period, 19, 25, 26, 35, 66, 69, 96, 98, 107, 131, 137, 139, 141

Kaisen, Peter, *24*, *25*, 33, 56, 92, 95, 106, 107, 108, 109
Kan Chuen Pao, 57, 112
Kansas, 38, 139
Kelly, Jeanne, 89

La Brea tar pits (California), 44, 115
Laramie Formation (South Dakota), 34
Life of a Fossil Hunter (Sternberg), 100, 134
Loomis, F. B., 44
Lystrosaurus, 71

Macrauchenia, 128
Mader, Bryn, 75
Maisey, John G., 129
Mammals, 18, 28, 31, 35, 36, 43, 44, 50, 57, 60, 67, 71, 74, 75, 77, 81, 112–128
Mammoths, 9, 43, 115
Mammut, 120–121, *121*
Mammuths, 122–123, *122*
Mammuthus, *115*
Marsh, Othniel Charles, 18, 23, 25, 33, 96, 107, 109, 139
Mastodons, 9, 43, 60, 74, 115, 126
Warren, 19, 120, *121*
Matthew, W. D., 7, 19, 33, 36, 40, 42, 52, 60, 65, 66, 67, 71, 92, 101, 103, 104, 105, 106, 112, 115, 123, 127, 140
McIntosh, John, 96
McKenna, Malcolm, 11, 71, 77, 79, 82, 113
McKenna, Priscilla Coffey, 79, 81
Megaloceras, *90–91*, 105, 123, *123*

Megalocnus, 28, 31

Merychippus, 44–45

Mesohippus, 41

Mesonychidae, 112, 113

Mesonyx, 112

Miocene period, 44, 88, 117, 119, 120, 126, 127

Missouri Breaks, 28

Mongolia, 10, 74, 77–83, 88, 103, 106, 113, 125

Monoclonius, 92

Mononykus, 10, *10*, 11, 80, 82, *83*, 88

Montana, 25, 28, *28*, 36, 44, 77, 93, 94, 103, 108, 109, 139

Morgan, J. P., 19, 26, 50, 120

Moropus, *44–45*, 45, *88*

Morris, Frederick K., 50, 57

Morrison Formation, 69

Moschops, 124, *124*

Mott Creek (Montana), 108

Nebraska, 44–45, 127

Neohipparion, 42

The New Conquest of Central Asia (Andrews), 126

New Mexico, 36, 40, 50, 62–63, 65, 68, 69, 99, 104

New York, 120

Norell, Mark, 11, 79, 80, 82, 83, 88

Notharctus, 11, 76, *76*, 77

Novacek, Michael, 79

Oligocene period, 113, 114, 125

Olsen, George, 56, 57, 59, *59*, 103, 106

Omnivores, 115

On the Origin of Species (Darwin), 98

Oreodonts, 75

Ornitholestes, 26, *27*

Ornithomimus, 92

Ornithopods, 103

Orohippus, 41

Osborn, Henry Fairfield, 10, 17–21, *19*, *21*, 24, 28, 31, 33, 36, 38, 40, 41, 42, 44, 45, 48, 49, 60, 64, 71, 72, 75, 82, 88, 95, 100–101, 104, 112, 113,

114, 118, 120, 127, 134, 140

Ouranosaurus, 117

Oviraptor, 59, *59*, *102*, 103

Pachycephalosaurus, 103, *103*, 124

Pachyrukhos, 28

Panthera leo atrox, 115

Paraceratherium, 10, 52, *52*, *53*, 74, 125–126, *125*

Parahippus, 41

Patagonia, 24, 28, 67, 68

Pelycosaurs, 116, 117

Pentaceratops, 40, 104–105, *104*

Permian period, 40, 116, 124, 133, 134

Phytosaurs, 140

Pinacosaurus, 59

Pinnipeds, 117

Plateosaurus, 105, *105*, 136, *136*

Platybelodon, 60, 126–127

Pleistocene period, 28, 31, 35, 43, 115, 118, 120, 122, 123, 128, 130

Plesiosaurs, 117, 131

Pliohippus, 42

Pliosaurs, 117

Porpoises, 119

Posen, Melissa, 88

Powder Hill (Montana), 103

Predators, 26, 59, 93, 99, 103

Preparation and mounting, 88–91

Primates, 11, 76, 77

Proganochelys, 136, *136*

Propalaehoplophorus, 28

Prosauropods, 105

Protoceratops, *58*, 59, 60, 74, 81, *81*, 103, 106, *106*, 140

Protohippus, 41

Protosuchus, 137, *137*

Psittacosaurus, 106

Pteranodon, *138*, 139, 140

Pterosaurs, 139

Quetzalcoatlus, 139

Red Deer River (Alberta), 32–33, *32*, 40, 44, 92, 95

Reptiles, 117, 124, 131, 133, 139, 140

Rhamphorhynchus, 139, *139*

Rhinoceroses, 10, 44, *44–45*, 45, 49, 52

Roosevelt, Theodore, 42

Ruminants, 117

Ruth Hall Museum of Paleontology, 69, 71

Rutiodon, 140, *140*

Sabertooth cats, 115

Santana Formation, 129

Saurolophus, 24, 33, *33*

Sauropods, 25, 26, 64–67, 96, 105, *105*, 136

Saurornithoides, 59

Scarritt Expeditions, 67

Schaeffer, Bobb, 99, 139

Shabarakh Usu (Mongolia), 52, 56–60, 72, 81, 103, 106

Shackelford, J. B., 50, 52, 125

Sicard, E. G., 65, 66

Simpson, George Gaylord, 21, 25, 42, 50, 63, 65, 66, 67, 114, 126

Sloths, 28, 31, 115

Smilodon, 115, 128

Sorenson, Carl, 99, 139

South America, 67–68, 128, 129

South Dakota, 42

Specimen types, 59

Spinosaurus, 117

Stegosaurus, 18, 107, *107*

Stenomylus, 44, *44–45*, 127, *127*

Stenopterygius, 141, *141*

Sternberg, Charles H., 33, 36, 38–40, *38*, *39*, 42, 92, 100, 104, 105, 109, 134

Sternberg, George, *38*, 100

Sternberg, Levi, 100

Styracosaurus, 33, *33*

Tedford, Richard, 42, 71, 72, 74

Tenontosaurus, 108, *108*

Texas, 36, 38, 40, 42, 64–67, 116, 117, 118, 133, 134, 139

Theropods, 82, 92–93, 98, *98*, 99, *99*, 103, 108, 109

This Business of Exploring (Andrews), 47, 48

Thomson, Albert "Bill," 19, 36, 42, 44, 45, 101, 126

Toxodon, 128, *128*

Trachodon, 38, 100, 101

Triassic period, 68, 71, 99, 105, 136, 137, 139, 140

Triceratops, 9, 18, 24, 28, *29*, 34, *39*, 93, 100, 103, 104, 109, *109*

Turtles, 77, 129, 130, 136, *136*

Type specimens, 75

Tyrannosaurus, 9, *9*, 10, 11, *11*, *22–23*, 24, 28, *30*, 31, 74, 88, 89, *89*, 93, 96, 109, 140

Ungulates, 119

"Unified Theory of Religion and Evolution" (Simpson), 42

Utah, 50

Varanosaurus, 117

Velociraptor, 59, 60, 81, *81*, 82

Walsh, Jeremiah, *24*

Warren, John Collier, 120

Warren Collection, 19

Warsavage, Steve, 89

Whales, 48, 117, 119

Whitaker, George O., 69, 99

Whitney, William C., 41

Williston, Samuel, 23

Winkley, William, 103

Wortman, Jacob L., 19, *25*, 26, 28

Wyoming, *21*, 25–27, 36, 44, 50, 76, 77, 96, 107, 109

Zalambdalestes, 81

ILLUSTRATION CREDITS

©American Museum of Natural History: Charles Knight/AMNH: Pages 8-9; 53; 121; AMNH: Page 118; ©Matthew Bergman: Page 49 (map) ©Mick Ellison/AMNH 1993: Page 83 center; Field Museum of Natural History, Neg. #CK19TA, Chicago: Page 44-45 bottom; ©Ed Heck/AMNI 1993: Page 10 top; ©Frank Ippolito 1992: Pages 27 center; 111; 136; ©Michael Rothman 1993: Pages 2-3; 6; 30 bottom right; 41; 83 top; 92; 93 left; 97; 100; 104; 105; 107; 108; 109; 110; 112; 115; 116; 117; 124; 129; 131; 135; 137; Illustrations by Siena Artworks/ London, England, ©Michael Friedman Publishing Group 1994: Pages 59; 94; 95; 99; 102; 103; 106; 108; 113; 114; 119; 125; 126; 127; 128; 133; 138-139; 140; 141; Courtesy of the Smithsonian Institution: Pages 29; 122

Front and back jacket illustrations by Michael Rothman on Australian and UK editions only

Photo credits on copyright page